Just Transitions

Series Editor

Raphael J. Heffron, University of Pau and Pays de l'Adour,
Pau, France

Ezio Costa Cordella
Editor

The Ecological Transition as a Challenge to Law

palgrave
macmillan

Editor
Ezio Costa Cordella
Law School
University of Chile
Santiago, Chile

ISSN 2731-6041 ISSN 2731-605X (electronic)
Just Transitions
ISBN 978-3-032-22054-7 ISBN 978-3-032-22055-4 (eBook)
https://doi.org/10.1007/978-3-032-22055-4

The original submitted manuscript has been translated into English. The translation was done using artificial intelligence. A subsequent revision was performed by the author(s) to further refine the work and to ensure that the translation is appropriate concerning content and scientific correctness. It may, however, read stylistically different from a conventional translation.

This Palgrave Macmillan imprint is published by the registered company Springer Nature Switzerland AG.
The registered company address is: Gewerbestrasse 11, 6330 Cham, Switzerland

Acknowledgments This book was prepared within the framework of the Fondecyt Initiation research project "Tensions and Challenges of Law and Environmental Regulation in the Context of the Ecological Transition" (ID 11231233, 2023–2026), funded by ANID Chile.

I am grateful for the support of Sofía Rivera Berkhoff and Inés Racionero Gómez in reviewing this text.

Contents

List of Contributors

Joaquín Abarzúa Varela University of Chile, Santiago, Chile

Ezio Costa Cordella Law School, University of Chile, Santiago, Chile

José Miguel Fuentes Zuleta Faculty of Architecture, Design and Urban Studies, Pontifical Catholic University of Chile, Santiago, Chile; Nucleo Patrimonios – NupatS, Santiago, Chile

Ignacia Godoy Tello University of Chile, Santiago, Chile

Iván Ojeda-Pereira Department of Social Sciences, LUT School of Engineering Sciences, LUT University, Lappeenranta, Finland; Institute of Sociology, Pontificia Universidad Católica de Chile, Santiago, Chile

Felipe Pino Zúñiga University of Chile, Santiago, Chile

Dominga Planella Durruty University of Chile, Santiago, Chile

Sofía Rivera Berkhoff University of Chile, Santiago, Chile; Heidelberg University, Heidelberg, Germany

Manuel Tironi Institute of Sociology, Pontificia Universidad Católica de Chile, Santiago, Chile; Institute for Sustainable Development, Pontificia Universidad Católica de Chile, Santiago, Chile

CHAPTER 1

Flâneurs of Disaster? Ecological Transition, Justice, and Law

Ezio Costa Cordella

In a drop of water
the birds quench their thirst
they cool off
they look at themselves.
We must transform ourselves
they say to each other.
Once we were dinosaurs.

Elvira Hernández, "In a Drop of Water" (Hernández, 2013)

1 Understanding the Great Ecological Transition and Associated Transitions

Being in transition is a very common state of affairs when viewed on an appropriate timescale, despite the rigidity of some of our conceptual frameworks. Things change over time; they evolve; they die. A system is reordered by an external disturbance; it shifts and causes

E. Costa Cordella (✉)
Law School, University of Chile, Santiago, Chile
e-mail: ecosta@derecho.uchile.cl

E. Costa Cordella (ed.), *The Ecological Transition as a Challenge to Law*, Just Transitions,
https://doi.org/10.1007/978-3-032-22055-4_1

another system to change in turn; this is the normal state of material existence. Living beings adapt and change, or become extinct, giving rise to new forms of life. Matter and energy continue their eternal dance.

Why, then, is it important to examine the phenomenon of ecological transition? First, not all transitions are identical or involve the same phenomena; second, not all transitions occur within time frames that allow us to assimilate and integrate them into our ways of life; and third, because the current omnipotent pretension of humanity has set us on a path to disaster by refusing to acknowledge the fragility of things and by believing that through a profound act of self-persuasion, we can forget it and create absolute stability and certainty for ourselves.

The ecological transition we are currently experiencing is characterized by its artificially (human-induced) accelerated pace and therefore does not align with other natural rhythms that would allow for reasonable adaptation. With a rise in global temperature occurring over just a few decades, there is no possibility for most living beings to adapt to the new conditions of their habitats, and even less for social systems to react in time to a series of changes they cannot even foresee.

To understand a transition means, first, to recognize that systems are embedded in an external environment and remain open to its influences, and then to recognize that variations in that environment generate changes within the system under analysis. Finally, a typical feature of transition processes is that the system reacts in a nonlinear manner as it adjusts to its context, transforming its own structures (Loorbach, 2009).

The close dependence of social systems on ecosystems makes this phenomenon even more evident, commonly referred to as *ecological transition*. This is understood as "the process of transformation during which a system shifts from one equilibrium regime to another, (...) a fundamental reconfiguration of the functioning and organization of the system" (Boissonade, 2017: 3). This conception aligns with the broader notion of transition as "a gradual process by which unsustainable structures are dismantled and new structures are built according to a specific design for sustainability" (Tremblay, 2011: 11).

Given the complexity of changes in the natural system and in various social systems and subsystems, what we are facing can be categorized as a major transition, but it can also be examined through the lens of a particular area. For example, we may refer to a water transition when discussing changes in water systems resulting from climate change; this, in turn, can be territorially situated (e.g., the water transition in the Nile Basin) or technically situated (e.g., the water transition in urban sanitation systems), or both simultaneously.

As we argued in previous work, ecological transitions can be analyzed, first and foremost, according to their intentionality. There are intentional transitions, in which social or institutional actors deliberately seek to transform a system toward a new equilibrium, and unintentional transitions, which occur as a result of ecological or social processes that were not explicitly directed by anyone. In this sense, a region may deliberately initiate a process of productive transformation—constituting an intentional transition—while phenomena such as desertification resulting from water overexploitation correspond to unintentional transitions, even if their causes are anthropogenic. The distinction lies not in the origin of the change but in the intentionality of the outcomes pursued (Costa, 2025).

Furthermore, transitions can be classified as governed or ungoverned, a category independent of their intentionality. A governed transition is one in which public policies, regulations, institutions, or instruments are designed to steer its effects toward specific social objectives; in contrast, an ungoverned transition occurs organically in an organic manner, without institutional direction to channel its impacts. Thus, the energy transition is an example of an intentional and governed process, whereas the water transition faced by many countries, driven by climate change, is unintentional, although it may eventually become governed if appropriate instruments are implemented. This distinction is central, as it determines when and through what mechanisms it is possible to incorporate justice criteria into the transitional process (Costa, 2025).

Moreover, the speed of the transition, its scale, and the uncertainty surrounding its contours have made it difficult to comprehend as a complex material phenomenon and have instead led to it being viewed from the various domains in which it occurs, in a relatively isolated manner.

2 Social Systems in Transition

In this volume, we are particularly interested in researching how law, as a system of norms and principles aimed at regulating society under an idea of justice and order, relates to the ecological transition. This involves following two different paths. The first concerns the way in which law influences, or should influence, transitions in various areas. Recognizing that ecosocial or sociotechnical transitions will be occurring in countless systems, a normal legal response would be to incorporate rules that allow these transitions to take place in a manner that maintains the ideas of order and justice that underpin the law.

So far, the law seems to have attempted this through two avenues: one more specific and regulated, manifested in the international climate regime and national climate change laws, and a second, more general and abstract, which is especially evident in the principle of a *just transition*.

2.1 Shepherding the Transition: Law as a Guide

The construction of law is rarely an orderly process, as it often takes place in pre-institutional spaces whose internal regulations are not necessarily protected by law. It is these sets of discourses, public policy debates, and material and economic realities that will shape the subsequent legal construction to institutionally channel collective preferences in a language that organizes actions. In the case of regulation concerning the variables of the ecological transition, part of the discussion that has become law has focused on the idea of a just transition. It is precisely this epithet of "just" that marks the insertion of law into the sociotechnical and ecosocial process of transition.

To observe this phenomenon, Iván Ojeda and Manuel Tironi, in their chapter "From the Global Just Transition to a Just Socioecological Transition in Chile? Trajectories and Technopolitics of Justice" show that the notion of a just transition emerges not as a stable or univocal category but as a technopolitical project in constant dispute, whose historical trajectory is marked by appropriations, reinterpretations, and struggles over meaning. Their reconstruction demonstrates that this category emerges from North American trade unionism and then moves into international climate debates, progressively expanding to new scales, actors, and normative contents. In this process, justice ceases to be understood solely as labor protection and becomes a guiding principle for broader sociotechnical transformations, opening a space for law to be challenged by demands that go beyond traditional sectoral or environmental regulation (Ojeda & Tironi, 2026).

On the basis of this genealogy, the authors argue that the just transition now operates as a heterogeneous assemblage capable of articulating labor, climate, and territorial concerns, and that its deployment in different contexts reflects tensions between the global and the local. This dynamic is particularly relevant for understanding the Chilean case, where the concept begins to be reconfigured by social and environmental actors, and subsequently is translated institutionally under the idea of a *just socio-ecological transition*. In this process of translation—marked by frictions, adaptations, and new ethical–political horizons—it becomes evident how transitional justice serves as a vector for state intervention and, at the same time, as a contested space regarding the scope and limits of public action in scenarios of ecosocial change.

On the other hand, José Miguel Fuentes, in "The Ecological Transition and the Green Deal: A Review of the (Counter)currents Toward a Just Transition," compares the discourses seeking to influence the concept of just transition, examining how some have been more successfully incorporated into public policy than others. The analysis reveals how the notion of just transition is also strained by a set of discourses aiming to shape its political and regulatory orientation. The author notes that in response to the advance of the Green Deals, countercurrents have emerged that criticize the way these programs seek to drive the ecological transition without altering the foundations of capitalism. These currents—including degrowth, post-growth, ecofeminism, and ecosocialism—warn that dominant institutional strategies are limited to promoting a technological and financial "big push" that leaves the extractivist logic and dependence on economic growth intact. Transition policies, therefore, are not only technically complex but also the subject of contestation among normative projects that offer divergent definitions of what should be understood as a truly just transition.

Drawing on these countercurrents, Fuentes demonstrates that the just transition must be understood as a field of controversies in which different frameworks of justice compete to guide public action. While the Green New Deal advances a reformist vision seeking to reconcile growth, ecological modernization, and the reduction of inequalities, critical discourses argue that without structural transformations in consumption patterns, labor organization, the distribution of economic power, and the relationship with nature, any transition risks becoming a form of *greenwashing*. This debate highlights the role of law as a guide for the transition: far from merely implementing technical decisions, it must choose between models of justice that are not equivalent and that chart divergent political horizons, making it all the more necessary to make explicit the normative premises underpinning any regulation aimed at shepherding the ecological transition (Fuentes, 2026).

Moreover, this concept of just transition has already been incorporated in various ways into law and public policy, a matter comparatively analyzed in "The Just Transition Paradigm and Its Comparative Expression" by Joaquín Abarzúa. The author shows that once the concept of just transition enters the legal field and public policy, it acquires specific configurations that respond both to the institutional characteristics of each country and to the available interpretive frameworks. His comparative analysis reveals that despite the polysemy identified in the literature, the various normative expressions of the just transition share a common core of distributive, procedural, restorative, and recognition principles, allowing us to speak of a single paradigm rather than multiple disconnected models. This conceptual unity is evident in the fact that the different typologies (labor, energy, socioecological) operate as variants of the same effort aimed at reconciling decarbonization, social justice, and inclusion.

In examining the nationally determined contributions (NDCs), biennial transparency reports (BTRs), and climate change laws of different states, Abarzúa finds that these instruments tend to institutionalize measures that essentially seek to mitigate the social costs of the transition and expand the collective benefits of economic transformation processes.

Additionally, the comparative review reveals that in practice, the most consolidated expressions of the just transition paradigm correspond to the labor and energy orientations, while the socioecological dimension still emerges only marginally. This asymmetry reflects not only states' preferences for "managerial" approaches but also the limits of what is legally possible within institutional frameworks that seek to manage the transitions within the existing economic model. In this regard, Abarzúa highlights a key point: the just transition is not only a normative ideal but also an operational category that already structures regulatory agendas, defines instruments, distributes competencies, and guides state action in climate governance. In this sense, his work demonstrates that law not only shepherds the transition but also helps stabilize a shared imaginary that integrates justice, sustainability, and equity as minimum parameters of legitimacy in contemporary ecological transformation (Abarzúa, 2026).

2.2 The Ecological Transition and the Law

Just as the law seeks to introduce order and justice into transitions, it must also undergo its own transformation and find ways to achieve balance with a material world that has lost the certainties that once formed the basis of our social agreements. The problem we face concerns our social expectations and the way we have constructed them, starting from the idea of stabilizing our existence through the so-called social contract, an abstract and ideal element that would contain the tacit agreement of all to generate an order allowing us to escape a state of nature that is chaotic and painful (Hobbes, 1980).

Indeed, the relative stability of nature and its cycles underpins our society and institutions, ensuring predictability and order. This stability has long been considered essential for the social contract and the functioning of the public sphere. The rhythms of the seasons, the reliability of the water cycle, and dependence on ecosystems are fundamental to our communities. Without such constancy, the foundations of social structures are compromised and the stability of a social agreement is shaken. In this volume, Dominga Planella analyzes the issue by arguing that the climate and ecological crisis could be considered a new *state of nature*, leading us to the need to reconfigure a significant part of our social contract.

In her chapter, Planella contends that the contemporary environmental crisis not only challenges the material foundations of the modern social contract but also strains its deepest normative premises. Drawing on her reconstruction of contractarian theory, the author argues that the triple planetary crisis operates as a structural phenomenon that reproduces the conditions of the state of nature: insecurity, lack of protection, and absence of effective guarantees. From this perspective, the collapse of ecological balances calls into question the state's ability to fulfill the sovereign function assigned to it by Hobbes, Locke, and Rousseau—namely, to ensure the life, liberty, and material stability of the political community. The erosion of these pillars, according to Planella, reactivates the very logic underlying the social pact: when authority ceases to provide sufficient protection, individuals find themselves once again in a precontractual space, marked by fear, threat, and a pervasive sense of vulnerability.

On this basis, the author suggests that the ecological transition can be understood as the intermediate movement between this new state of nature induced by the crisis and the possible formulation of a renewed social contract. The transition thus appears not as a merely technical process but as a period of instability in which systems are pressured to adapt or transform. As she notes, this stage of historical uncertainty requires asking which elements of the modern order can be preserved and which must be changed in order to comply with planetary limits, bearing in mind that while change is inevitable, its direction is a matter of choice. In this sense, her analysis invites us to understand the ecological transition as a decisive political space in which the content of the new pact is contested: one that recognizes biophysical limits, reconfigures the relationship between society and nature, and restores minimum conditions of material security for communal life (Planella, 2026).

This issue is highly relevant because the very legitimacy of the rule of law depends on its ability to provide certain minimums, among which physical security, now increasingly linked to the risks posed by the climate crisis, is paramount.

In this regard, the scientific literature shows that the climate crisis directly threatens the state's capacity to guarantee physical, health, and economic security—functions essential to the viability of the social contract. For example, the Intergovernmental Panel on Climate Change (IPCC) warns that the increase in heat waves raises mortality, reduces labor productivity, and severely affects human health, especially among vulnerable groups, thereby undermining the most basic duty of preservation that underpins political authority (IPCC, 2023, pp. 1041–1170). Added to this is evidence from the European Environment Agency, which projects significant losses in critical infrastructure and essential systems such as energy, water, and transport because of increased droughts, floods, and storms, affecting the continuity of services that sustain daily life and social cohesion (EEA, 2024, p. 11, 2025).

These documented risks simultaneously erode well-being, material certainties, and expectations of protection—central pillars of any legitimate social pact. At the same time, they deepen existing inequalities and generate new forms of vulnerability, undermining the social justice that sustains the legitimacy of the legal order. Botzen et al. (2020) show how lower-income households face greater economic losses from extreme events because of their reduced adaptive capacity, reproducing patterns of distributive injustice that the social contract seeks precisely to correct.

In the same vein, international courts have stated that the climate crisis not only threatens environmental goods but also destabilizes the very foundations of social coexistence. The International Court of Justice, in its Advisory Opinion on the obligations of states regarding climate change, expressly recognizes that global warming constitutes an "urgent and existential threat" to humanity and to the life systems on which societies depend, highlighting that phenomena such as the loss of habitable territory, food insecurity, and population displacement undermine the material conditions for the exercise of human rights and the validity of the legal order (ICJ, 2025, pp. 13–15). The Court concludes that states have legal obligations—and not merely political aspirations—to prevent significant harm, cooperate internationally, and adopt mitigation and adaptation measures guided by the best available science, precisely to avoid scenarios of environmental degradation that could lead to crises of legitimacy and the breakdown of the social contract (ICJ, 2025).

Meanwhile, the International Tribunal for the Law of the Sea has emphasized the risks to social stability arising from the degradation of the oceans. In its 2024 Advisory Opinion, issued at the request of the Commission of Small Island States on Climate Change and International Law, the Tribunal held that greenhouse gas emissions absorbed by the ocean constitute "marine pollution" under the United Nations Convention on the Law of the Sea, and that States Parties are obliged to prevent, reduce, and control such emissions diligently, using the best available science and increasingly stringent international standards (ITLOS, 2024, paras. 491–499). By emphasizing the particular vulnerability of small island States to sea level rise, ocean acidification, and warming—with direct impacts on fisheries, food security, the habitability of coastal territories, and the cultural continuity of communities—the Tribunal explicitly links marine environmental degradation to risks of economic disintegration, social conflict, and loss of community cohesion across entire regions (ITLOS, 2024).

Finally, the Inter-American Court of Human Rights, in its Advisory Opinion OC-32/25 on climate emergency and human rights, goes even further by identifying the climate crisis as a factor of structural instability for the societies of the continent. The Court asserts that states have the obligation to prevent, mitigate, and remedy the impacts of climate change because these directly threaten rights such as life, personal integrity, health, water, food, housing, and political participation, emphasizing that phenomena such as heat waves, prolonged droughts, floods, and rising sea levels can lead to forced displacement, loss of livelihoods, and deepening poverty (Inter-American Court of Human Rights, 2025, paras. 66–88, 358–367). By requiring States to adopt adaptation measures and ecological transition policies with a human rights approach, the Court warns that climate inaction not only constitutes a violation of conventional obligations but also increases the risk of socioenvironmental conflicts, erodes trust in institutions, and ultimately calls into question the very stability of democracy in the Americas (Inter-American Court of Human Rights, 2025).

Sofía Rivera's proposal is particularly relevant because it allows us to understand that addressing the ecological risks threatening social stability requires first transforming the way we think about law. Her ecological analysis of law begins by diagnosing that the modern legal paradigm—based on the human/nature dualism, the centrality of individual freedom, and the mechanistic view of legal phenomena—not only is incapable of responding to the climate crisis but also constitutes one of its structural enabling conditions (Rivera, 2026).

Within this framework, she argues that the modern separation between law and the natural world prevents recognition of the ecological interdependencies that sustain any form of social organization, reproducing relations of domination, rendering communities invisible, and promoting legal institutions that reinforce both environmental exploitation and historical social inequalities. Therefore, an ecological interpretation of law is not limited to introducing environmental considerations into legal doctrine but also requires reconfiguring the philosophical foundations of law by incorporating systems theory, the notion of complexity, and a material understanding of the biophysical limits that condition all communal life (Rivera, 2026).

In this sense, and given the search for balance characteristic of periods of transition, Rivera's proposal acquires decisive relevance: she suggests that law must abandon the modern aspiration for certainty and mechanical order, replacing it with a conceptual framework that recognizes the dynamic, uncertain, and relational character of the Earth system. This shift would allow law to adopt methodologies and principles that integrate contingency, the openness of the legal system, and the need to manage the relationship between society and environment in a balanced manner (Rivera, 2026).

From the author's perspective, this enables not only a better understanding of the socioecological conflicts that are already destabilizing entire communities, but also the construction of institutions capable of governing the ecological transition without reproducing the injustices of the modern paradigm. Thus, the ecological analysis of law provides a road map for rethinking freedom, property, authority, and community according to parameters compatible with the continuity of life, restoring to law its role as guarantor of coexistence on a planet marked by limits and profound interdependencies.

3 Niches in Transition

Given the existence of ongoing transition processes, observing their practical application can be highly relevant for appreciating their contours and for understanding the extent to which law can play, or has played, a role.

A first practical approach to niches in transition can be observed in the field of private sector green finance, where, as Ignacia Godoy shows, a set of instruments is emerging that are beginning to play a strategic role in directing investments toward environmentally sustainable activities. The author demonstrates that green bonds, bilateral green loans, and sustainability-linked financing constitute mechanisms that function as intermediate spaces between the traditional financial structure and the objectives of the ecological transition, enabling companies of various sizes to access capital for mitigation or adaptation projects (Godoy, 2026).

These instruments, guided by voluntary principles such as the *Green Bond Principles* or the *Green Loan Principles*, are configured as normative niches where law still operates in a fragmented manner, but in which standards of transparency, reporting, and verification are being tested that could eventually be consolidated as legal obligations. In this sense, the financial market appears not only as a passive recipient of regulations but also as a laboratory where new practices are incubated that challenge, complement, or even anticipate state regulatory frameworks.

However, these niches also reveal tensions that illustrate the current limits of law in steering the ecological transition. Godoy notes that the absence of uniform definitions of what constitutes a "green" investment, as well as the lack of binding regulatory frameworks, creates space for *greenwashing* practices that can undermine the transformative role of sustainable finance (Godoy, 2026).

Furthermore, the author identifies that unequal access to green finance—especially for small and medium-sized enterprises or for projects located in countries with limited institutional capacities—creates niches where the transition advances asymmetrically, benefiting certain sectors or territories, while others are left behind. In these spaces, law has the opportunity to intervene to establish minimum standards, mandate transparency, and ensure that the benefits of the transition are not captured exclusively by large financial actors.

In turn, Pino's analysis allows us to observe that some of the most dynamic niches of the energy transition are emerging precisely at the intersection of global markets, national policies, and the expectations of private actors. The chapter demonstrates that although International Energy Law has been envisioned as a regime capable of guiding the transition, in practice many of these niches develop through hybrid arrangements that incorporate trade agreements, cooperation mechanisms, public–private partnerships, and technical standards that advance more rapidly than the consolidation of the legal regime itself. This situation creates zones of regulatory innovation where Latin American states seek to strategically position themselves to attract investment related to renewable energy, "critical minerals," or green hydrogen, even though these strategies do not always translate into internal transformations of their energy matrices. As Pino notes, this generates spaces where law functions more as a facilitator of global economic–technological agendas than as a driver of structural transformations in sustainability and equity (Pino, 2026).

At the same time, these niches reveal both the limits and opportunities for law to influence ongoing transition processes. Pino shows that in Latin America, the rapid expansion of agendas such as the exploitation of critical minerals or the establishment of green hydrogen hubs has led to a proliferation of regulatory and public policy instruments aimed at positioning states as strategic suppliers in the global decarbonization chain. However, when these instruments are primarily focused on ensuring international competitiveness, without clear redistribution mechanisms or socioenvironmental safeguards, there is a risk of reproducing extractivist patterns that undermine the legitimacy of the transition process itself. At this point, the niches function as regulatory laboratories where the capacity of law to balance economic incentives with environmental protection, public participation, and energy justice standards is put to the test. As the author suggests, it is precisely in these spaces that one can assess the extent to which law has managed to play a leading role, and not merely be a passive enabler, in a just energy transition in the region (Pino, 2026).

4 Not All Who Wander are Lost

A significant part of the risk associated with the ecological transition lies in the possibility that its speed and depth may exceed or even nullify the adaptive capacities of those living today and of future generations. We are living through a historical moment in which environmental variables have lost the stability that for centuries served as the foundation of our institutions, shifting toward a new equilibrium whose contours remain unknown. The only certainty is that unless we halt the causes of the climate and ecological crisis, the path ahead will be marked by increasingly frequent and intense disasters.

In the face of this uncertain scenario, it becomes essential to establish minimum principles to guide the transition with as little harm as possible, a task for which law is indispensable. However, the rule of law itself is being eroded in this transition, unable to fully grasp the material conditions that underpin its foundations. Our social order lacks adequate rules to guide the multiple ongoing processes; people are being harmed, and these deeply unjust harms remain unaddressed, revealing a troubling impotence of law to deliver justice.

The detachment with which law has so far responded to the ecological transition is costing it legitimacy and even continuity. For this reason, incorporating transitional logics is no longer merely desirable, but urgent. Navigating an unknown path entails difficulties, but wandering is not the same as being lost: the wanderer moves forward with a general sense of direction, guided by principles that prevent deviation from what is essential; the lost, by contrast, roam without a horizon. Law must choose the former.

This book seeks to illuminate certain points that may help imbue this transition with a sense of justice. For although law cannot provide us with an exact map, it can and must advance with the perseverance of the wanderer, and not with the cynical detachment of the *flâneur*—a detachment that abandons justice and opens the way to naked powers that deny human dignity. Only by avoiding such cynicism can we aspire for the ecological transition to also be a just transition.

References

Abarzúa, J. (2026). The just transition paradigm and its comparative expression. In E. Costa (Ed.), *The ecological transition as a challenge for Law*. Palgrave Macmillan.

Boissonade, L. (2017). *La Transition. Analyse d'un concept.* Thém [online]. Ministère de la Transition écologique et de la Cohésion des territoires. Available at https://www.ecologie.gouv.fr/sites/default/files/Th%C3%A9ma%20-%20Transition%20-%20Analyse%20d%27un%20concept.pdf. Accessed December 9, 2025.

Botzen, W. J. W., Deschenes, O., & Sanders, M. (2020). The economic impacts of natural disasters: A review of models and empirical studies. *Review of Environmental Economics and Policy*, *13*(2), 167–188.

Costa, E. (2025). The inevitable ecological transition and the desirable just transition. *Journal of the Master's in Applied Systems Analysis to Society*, *52*, 39–58.

EEA [European Environment Agency]. (2024). *Europe's state of water 2024: The need for improved water resilience.* Publications Office of the European Union. https://data.europa.eu/doi/10.2800/02236

EEA [European Environment Agency]. (2025). *Europe's environment 2025: Knowledge for resilience, prosperity and sustainability.* EEA.

Fuentes, J. M. (2026). The ecological transition and the green deal: A review of the (counter)currents toward a just transition. In E. Costa (Ed.), *The ecological transition as a challenge for Law*. Palgrave Macmillan.

Godoy, I. (2026). The role of green finance in the private sector as a tool for a just transition. In E. Costa (Ed.), *The ecological transition as a challenge for Law*. Palgrave Macmillan.

Hernández, E. (2013). *The flag of chile.* Cuarto Propio. (Cited poem: "In a drop of water").

Hobbes, T. (1980). *Leviathan.* Fondo de Cultura Económica. (Originally published in 1651).

ICJ [International Court of Justice]. (2025). Obligations of States in respect of climate change (Advisory Opinion of 23 July 2025). ICJ Reports.

Inter-American Court of Human Rights. (2025). *Advisory opinion OC-32/25: Climate emergency and human rights.*

IPCC [Intergovernmental Panel on Climate Change]. (2023). *Climate change 2023: Impacts, adaptation and vulnerability.* Contribution of working group II to the sixth assessment report of the intergovernmental panel on climate change. Cambridge University Press.

ITLOS [International Tribunal for the Law of the Sea]. (2024). *Request for an advisory opinion submitted by the commission of small Island States on climate change and International Law*. Advisory Opinion, Case No. 31.

Loorbach, D. (2009). Transition management for sustainable development: A prescriptive, complexity-based governance framework. *Governance. An International Journal of Policy and Administration*, *23*(1), 161–183. https://doi.org/10.1111/j.1468-0491.2009.01471.x

Ojeda, I., & Tironi, M. (2026). From the global just transition to the socio-ecological just transition in Chile? Trajectories and technopolitics of justice. In E. Costa (Ed.), *The ecological transition as a challenge for Law*. Palgrave Macmillan.

Pino, F. (2026). International energy law in the Latin American energy transition: Avoiding a race to the bottom. In E. Costa (Ed.), *The ecological transition as a challenge for Law*. Palgrave Macmillan.

Planella, D. (2026). The environmental crisis and social contract theory within the framework of the ecological transition. In E. Costa (Ed.), *The ecological transition as a challenge for Law*. Palgrave Macmillan.

Rivera, S. (2026). Ecological analysis of law: Proposal for a way of thinking about law for the ecological transition. In E. Costa (Ed.), *The ecological transition as a challenge for law*. Palgrave Macmillan.

Tremblay, L. (2011). *Governance of transitions towards sustainability*. Université de Sherbrooke.

CHAPTER 2

From the Global Just Transition to a Just Socioecological Transition in Chile? Trajectories and Technopolitics of Justice

Iván Ojeda-Pereira and Manuel Tironi

1 Introduction: Transition(s) and Justice(s)

In response to the emergence of the global socio-environmental catastrophe, various technopolitical projects have been articulated to address the scenario. Thus, different discursive fields have emerged around transitions

AI Use Statement: Artificial intelligence (AI) tools were used to support language editing and translation in the preparation of this manuscript. The author takes full responsibility for the content, interpretation, and originality of the work.

I. Ojeda-Pereira (✉)
Department of Social Sciences, LUT School of Engineering Sciences, LUT University, Lappeenranta, Finland;
Institute of Sociology, Pontificia Universidad Católica de Chile, Santiago, Chile
e-mail: ilojeda@uc.cl; Ivan.Ojeda-Pereira@lut.fi

M. Tironi
Institute of Sociology, Pontificia Universidad Católica de Chile, Santiago, Chile;
Institute for Sustainable Development, Pontificia Universidad Católica de Chile, Santiago, Chile

E. Costa Cordella (ed.), *The Ecological Transition as a Challenge to Law*, Just Transitions,
https://doi.org/10.1007/978-3-032-22055-4_2

(Pesch, 2015), each mobilising interests, objectives, and measures that reconfigure the "pathways" and morphologies of the very idea of transition (Geels & Schot, 2007). In this regard, it is essential to understand two characteristics of transitions. First, they are political and technical processes in construction, laden with histories, projections, exaggerations, and frictions (Ojeda-Pereira et al., 2024). Second, there are varieties of transitions.

The debate on what transitions are and how to govern them is extensive. More restricted perspectives argue that the challenge of transitions lies in the development of infrastructures and technologies for energy storage and operation, with an emphasis on technological efficiency and reliability (Chang et al., 2021). In contrast, broader approaches have defined sustainable transitions as sociotechnical processes oriented towards sustainability goals; the inclusion of this social component in the definition allows them to be analysed by the social sciences (Avelino et al., 2016).

In dialogue with these broader perspectives, the question of justice in this context of change has become consolidated in international debate over the past decade. This leads us to concrete questions: How can we advance transformation without generating new frameworks of exclusion or reproducing historically constituted forms of violence? What are we, as a society, willing to give up in the name of this change? What are the political, ethical, and economic limits of transitional interventions?

In this chapter, we analyse the construction and relationship of two ways of conceptualising Justice in the era of Transitions: the Just Transition and the Just Socioecological Transition. The Just Transition originated in the 1970s within North American trade unionism and has since evolved, reaching international scale and relevance (Morena et al., 2020). The Just Socioecological Transition approach has its roots in Chilean national environmental policies between 2010 and 2020—for example, in the Environmental and Social Recovery Programme (PRAS) initiatives and the first decarbonisation strategies led by the Ministry of Energy.

In this chapter, we will first describe the ethical and political content, proposing a periodisation of the social life of each of these approaches; second, we will reflect on and discuss the relationship between both modes of justice-oriented transition. We argue that, despite thematic connections and shared concerns, each approach has a unique social trajectory, shaped by diverse political and contextual content. Therefore, it would be reductive to understand the Just Socioecological Transition as merely a national implementation of the ideas of the Just Transition, but, at the same time, it would be a mistake to assume that the Just Socioecological Transition arises without any motivation from the Just Transition.

2 Three Global Phases of the Just Transition

In this section, we present a historical perspective on the Just Transition. In line with periodisation efforts (Morena et al., 2018), we propose three phases to understand its global circulation. In each of these stages, the Just Transition is adopted and appropriated by various actors, who shape and modify its content, normative ideals, and applications.

2.1 The Genesis of the Just Transition: Workers and the Environment

At its origin, the Just Transition had a direct connection with labour movements in the United States and Canada. Thus, the Just Transition carries a predominantly labour-oriented agenda, socially constructed at the local level and later scaling up to national levels. The political and social construction of the Just Transition can be traced back to the 1950s, inheriting the concerns of early environmentalists in the United States regarding nuclear weapons and the impacts of their testing (see Fig. 1).

In the 1960s and 1970s, the role of union leader Tony Mozzocchi of the Oil, Chemical and Atomic Workers Union in the United States was crucial in articulating the semantics of the Just Transition. Mozzocchi participated in the first Occupational Safety and Health Act in the United States and later, together with other unionists in 1973, organised the first environmental strike: Support America's. This protest brought together Shell refinery workers from different states across the United States over labour conditions, health, and workplace safety, which intersected with community demands regarding their own health and the environment (Gordon, 1998; Morena et al., 2018). Thus, the first environmental strike marked the beginning of an era in which, under union leadership, networks linking labour and environmental issues were created. The Oil, Chemical and Atomic Workers Union in the United States and the Communications, Energy and Paperworkers Union in Canada were central to this process.

In 1995, Les Leopold (from the Labor Institute) and Brian Kohler (a unionist from the Communications, Energy and Paperworkers Union of Canada), who had supported Mozzocchi 20 years earlier, used the concept of Just Transition in their presentation before the International Joint

United States 1950
So-called social environmentalism was promoted by organisations that opposed industrial capitalism, the development of nuclear weapons, and their testing in the atmosphere.

United States 1970
Tony Mazzocchi, a trade unionist from the Oil, Chemical and Atomic Workers Union, worked toward the approval of the Occupational Safety and Health Act.

United States 1973
Tony Mazzocchi, a trade unionist from the Oil, Chemical and Atomic Workers Union (OCAW, an organisation active between 1917 and 1999), gained support from environmentalists for the first environmental strike, "Support America's First Environmental Strike", related to health and safety issues at Shell refineries in four U.S. states. Mike Merrill and Les Leopold (Labor Institute), Brian Kohler (trade unionist from the Communications, Energy and Paperworkers Union of Canada), and Robert Wager (also from OCAW) supported this initiative.

United States and Canada 1993
The network of trade unionists led by Mazzocchi created the Superfund for Workers (Mazzocchi, 1993), an action programme that challenged the discourse of "jobs versus the environment." This opposition between environmental protection and employment was reinforced by studies funded by business groups and other trade union sectors; in general, these studies argued that stricter environmental regulation would negatively affect employment (for example, Gollop & Roberts, 1983).

United States and Canada 1995
Les Leopold (Labor Institute) and Brian Kohler (trade unionist from the Communications, Energy and Paperworkers Union of Canada) used the term Just Transition in their presentation to the International Joint Commission on Great Lakes Water Quality (Hampton, 2015).

Canada 1996
The Communications, Energy and Paperworkers Union of Canada adopted the Just Transition Resolution.

Fig. 1 Timeline of the genesis phase of the Just Transition. (Gollop & Roberts, 1983; Gordon, 1998; Hampton, 2015; ICFTU, 1997; Jenkins et al., 2020; Morena et al., 2018; Working Lives Research Institute, 2008)

Commission on Great Lakes Water Quality, thus marking its public emergence (Morena et al., 2018).

It is important to understand that business groups and some unions in the United States argued that improvements in environmental regulations would negatively affect job availability and quality, forging a technoeconomic narrative that the Just Transition seeks to challenge. Thus, the Just Transition emerged as a reaction to a political narrative that viewed environmental concerns as conflicting with employment opportunities (Ciplet, 2022).

Indeed, in this initial phase, the Just Transition was an emerging concept, politicised primarily by union actors in North America. The Just Transition argued that it is not change that will generate a crisis; rather, the crisis already exists in society, and change will allow us to overcome these historically accumulated problems. In this context, the Just Transition approach emerged as an invitation to social reflection, maintaining that employment is already precarious and that this is the real crisis, thus requiring change to improve workers' health and safety conditions, as well as environmental impacts (Gordon, 1998).

2.2 *Beyond Workers: On the Climatisation of the Just Transition*

Drawing on the conceptual proposal of Aykut and Maertens (2021) regarding the climatisation of global politics, in the second phase the Just Transition was refocused towards the climate crisis (see Fig. 2). This development was accompanied by a change in scale from North America to the arena of international environmental policy.

From the late 1990s until the end of 2015, the national–international link was strengthened, diversifying the actors who promoted the Just Transition and providing it with new contents and forms of existence. Thus, various organisations adopted resolutions on the Just Transition—for example, the International Federation of Chemical, Energy, Mine and General Workers Union and the International Trade Union Confederation (Morena et al., 2018). During this period, international unions, foundations, and nongovernmental organisations (NGOs) that were now engaging with the Just Transition also found resonance and built connections with different United Nations organisations. The concept began to be used by United Nations system bodies, such as the United Nations Research Institute for Social Development (Morena et al., 2018), the United Nations Global Compact (United Nations Global Compact, 2022), and the Conferences of the Parties to the United Nations Framework Convention on Climate Change (COPs) themselves (Stevis et al., 2020).

In this regard, it should be noted that supranational institutions began to play a more significant role. Thus, during this second phase, the North American political–labour influence shifted, incorporating new demands such as class, gender, and race (Stevis & Felli, 2016). We also observe that climate change emerged as a powerful driver

International 1997
The International Confederation of Free Trade Unions included the term Just Transition in its declaration at COP 3 in Kyoto: "Workers will demand an equitable distribution of costs through Just Transition policies that include measures for equitable recovery of the economic and social costs of climate change programmes" (ICFTU, 1997, p. 1).

United States 1997
The Oil, Chemical and Atomic Workers Union adopted its resolution on Just Transition.

United States 1997
The President of the American Federation of Labor and Congress of Industrial Organizations created the Blue/Green Working Group, a space for dialogue between environmental organisations and trade unions.

International 1997
The Just Transition Alliance (JTA) was created, connecting environmental activists, trade unions, and communities focused on environmental justice, and generating spaces for activism, education, and programmes.

Canada 1999
The Canadian Labour Congress adopted its own Just Transition Resolution.

International 1999
The International Federation of Chemical, Energy, Mine and General Workers' Unions adopted a Just Transition Resolution. Similarly, the International Confederation of Free Trade Unions, the Trade Union Advisory Council, and the Organisation for Economic Co-operation and Development followed.

Spain 2003
Workers' Commissions worked jointly with the Sustainlabour Foundation. Sustainlabour is a think tank that develops reports and trade union training programmes and is linked to international policy institutions such as the United Nations Environment Programme, the International Labour Organization, the United Nations Framework Convention on Climate Change, and the United Nations Commission on Sustainable Development.

International 2006
The International Confederation of Free Trade Unions joined with the World Confederation of Labour and created the International Trade Union Confederation.

United Kingdom 2005–2007
The Trade Union Congress funded a research project led by the Working Lives Research Institute, which resulted in the publication A Green and Fair Future: For a Just Transition to a Low Carbon Economy (TUC, 2008). Similar initiatives also emerged, guided by the Australian Council of Trade Unions.

International 2006
Sustainlabour organised the first Trade Union Assembly on Labour and the Environment.

Fig. 2 Two-part timeline tracing the international expansion and climatisation of the Just Transition concept between 1997 and 2015. *COP 3* Third Conference of the Parties to the United Nations Framework Convention on Climate Change, *Rio + 20 Conference* 2012 United Nations Conference on Sustainable Development. (Gollop & Roberts, 1983; Gordon, 1998; Hampton, 2015; ICFTU, 1997; Jenkins et al., 2020; Morena et al., 2018; Working Lives Research Institute, 2008)

International 2009
At the Copenhagen Conference, the International Trade Union Confederation presented Just Transition as "a tool the trade union movement shares with the international community, aimed at smoothing the shift towards a more sustainable society and providing hope for the capacity of a green economy to sustain decent jobs and livelihoods for all" (ITUC, 2009).

International 2009–2014
The United Nations Environment Programme, the International Labour Organization, the International Trade Union Confederation, and the International Organisation of Employers strengthened the Green Jobs Initiative, explicitly identifying the advancement towards a Just Transition as one of its objectives.

International 2012
The second Trade Union Assembly on Labour and the Environment was held on the occasion of the Rio+20 Conference.

International 2015
International trade unions made significant efforts to ensure that the language of Just Transition was adopted by United Nations agencies and programmes. The concept was included in the preamble of the Paris Agreement.

Fig. 2 (Coninued)

shaping the content of the Just Transition (Stevis et al., 2020); the crisis no longer was rooted only in historically accumulated injustice but also appeared as an adversary to be proactively confronted. Priority was no longer limited to labour and health legislation but shifted towards acting swiftly and effectively in response to the crisis, without repeating historical patterns of violence.

2.3 "We are All Part of Just Transition": Opening and Heterogenisation

From 2015 onward, in addition to maintaining its climate focus, the Just Transition became more heterogeneous through two developments. First, business actors adopted the discourse of the Just

Transition, resulting in a semantic broadening of the concept. Some of the most emblematic cases include the Just Transition Fund supported by the Rockefeller Family Fund and Chorus Foundation, and the launch of the partnership between the Just Transition Centre affiliated with the International Trade Union Confederation and the business groups B Team and We Mean Business, which developed a Just Transition Business Guide (Morena et al., 2018) (Fig. 3).

Second, the Just Transition became a guiding principle for national public policies. With the support of the United Nations Development Programme (UNDP) and its Climate Promise initiative, teams of Just Transition specialists began to be formed (UNDP, 2022). Typically, experts in social and political sciences prepare reports, operational frameworks, guidelines, and systematisations of public policy experiences.

International 2015
The Just Transition Fund was created, supported by the Rockefeller Family Fund and the Chorus Foundation. This event marked a milestone in the philanthropic diffusion of the concept. Subsequently, other funding sources such as Bloomberg Philanthropies, the European Climate Foundation, the Edge Funders Alliance, Building Equity, and Alignment for Impact, among other organisations, incorporated the language of Just Transition.

International 2016
The Just Transition Centre was launched, affiliated with the International Trade Union Confederation, inaugurating a new phase of trade union efforts. With a renewed perspective, the Centre collaborated directly with two business groups and produced a Just Transition Business Guide (The B Team and We Mean Business).

International 2018
During COP24 in Poland, the Solidarity and Just Transition Silesia Declaration was adopted.

Fig. 3 Timeline showing the diversification of Just Transition actors between 2015 and 2018, including philanthropic and business sector involvement. *COP24* 24th Conference of the Parties to the United Nations Framework Convention on Climate Change. (Gollop & Roberts, 1983; Gordon, 1998; Hampton, 2015; ICFTU, 1997; Jenkins et al., 2020; Morena et al., 2018; Working Lives Research Institute, 2008)

Under this funding, we find reports on the implementation of policies framed by the Just Transition discourse, addressing, for example, the cases of Serbia (Cavalheiro, 2020), Costa Rica (Groves et al., 2022), Zimbabwe (Svenja et al., 2021), Canada, Kenya, and Germany (Morena et al., 2018).

In this third phase, the content and scope of the Just Transition were reconfigured. Initially, the Just Transition emerged as a response to technoeconomic determinism that limited the debate on the environment and employment. Now, it appears as an entity whose normative horizon requires the design and implementation of public policy pathways. Furthermore, the Just Transition has become a mechanism linking trade unionists, environmentalists, supranational institutions, foundations, NGOs, business leaders, and philanthropists. This has inevitably led to an expansion of its field of meaning and, consequently, the emergence of controversies.

3 Four Phases of the Just Socioecological Transition in Chile

In this section, we present a historical perspective on the Just Socioecological Transition in Chile, organised into four phases: precursor mechanisms, genesis, public stabilisation, and outlook.

3.1 Precursor Political Mechanisms

Because of institutional and regulatory weakness, Chile has highly saturated extractive zones affected by industrial pollution. In this context, social and environmental movements imported the concept of sacrifice zones to designate these localities that were deliberately subjected to toxicity (Tironi, 2018). For some time, sacrifice zones were even viewed optimistically by public policy, as they represented progress and economic growth, and embodied the territorial clustering of industrial hubs in the Latin American Jaguar (Campos-Medina, Korneeva, et al., 2021; Campos-Medina, Ojeda Pereira, et al., 2021). However, as a result of persistent local criticism, the Chilean state recognised that these areas required remediation, and in 2014 it implemented the Environmental and Social Recovery Program (PRAS) (Pinto et al., 2023).

However, we consider these plans to have arrived at an inopportune time, belatedly; by this, we do not mean there was a "delay." On the contrary, we believe that at least in their formulation, they aimed to advance in a direction of social and ecological justice that was not yet the standard in Chilean public policy, and, for this reason, they were quickly relegated to a peripheral position within environmental public policy.

In accordance with the international agreements to which Chile is a signatory in the energy sector, after four years of implementation of the PRAS, the Ministry of Energy and the Ministry of the Environment under President Sebastián Piñera's administration convened a working group on decarbonisation. Thus, in 2020, Chile's Nationally Determined Contribution Report to the Paris Agreement incorporated the concept of the Just Transition into its social pillar (Government of Chile, 2020). This anticipated the results of the decarbonisation working group, which were published as the "Just Transition Strategy in the Energy Sector" (Ministry of Energy, Government of Chile, 2021). This public–private guide for the closure of coal-fired power plants emphasises the responsibility of both the state and private actors to address the challenges of labour reconversion.

From our perspective, the PRAS, Just Energy Transition reports, and collaborative workspaces emerged as experiments within various public institutions (see Fig. 4). Although they began to introduce the challenge of justice in the context of transition, they failed to coordinate efforts under an integrative approach. Hence, we propose to understand them as public policy mechanisms that paved the way for the incorporation of the Just Socioecological Transition.

3.2 *Genesis and Ethical–Political Debate: Justice in the Transition in Latin America?*

Starting in 2020, civil society organisations, NGOs, foundations, and research and study centres began to promote public discussion on justice in the context of the transition in Chile. Unlike the precursor mechanisms, this process began outside the state and succeeded in bringing together a greater number of actors, such as the Environmental Prosecutor's Office (FIMA) NGO, Fundación Chile Sustentable, and Fundación Espacio Público (see Fuentes et al., 2020; Lecourt et al., 2023; Rabi et al., 2021). All are national organisations with extensive experience in public environmental debates in Chile.

In this way, the ethical–political debate on how to integrate justice into the Latin American reality, and specifically in the Chilean context,

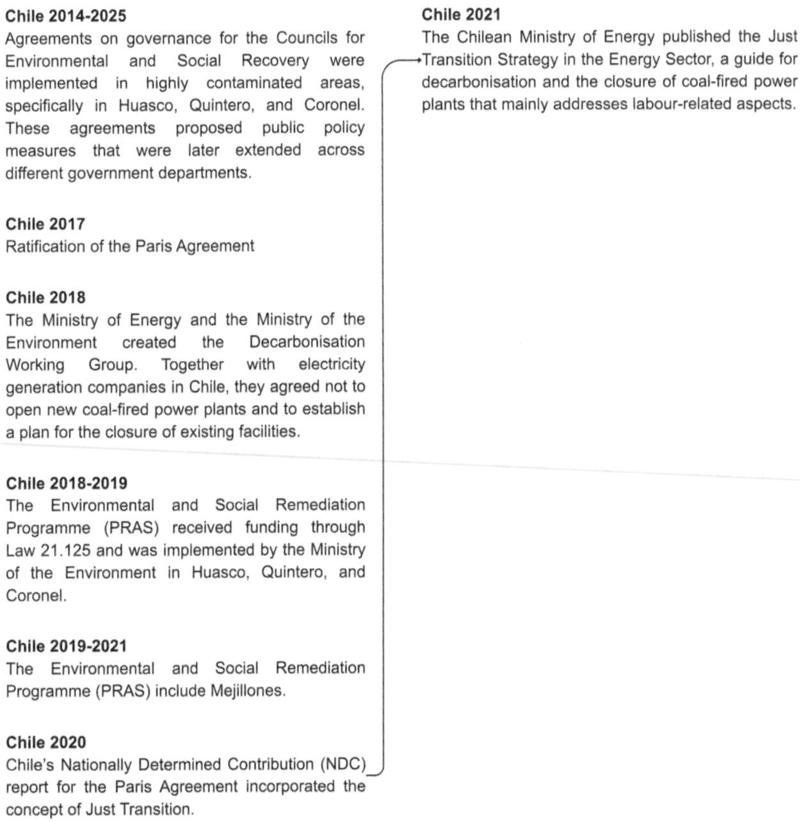

Fig. 4 Timeline of Chilean public policies from 2014 to 2021, showing increasing attention to decarbonisation and the Just Transition

took on various forms. In 2020, 2021, and 2023, collective projects were organised, resulting in reports, books, and non-academic publications with a particular focus on public outreach (see Fuentes et al., 2020; Lecourt et al., 2023; Rabi et al., 2021). In 2023, courses, seminars, and debates were also held, linking climate justice and the Just Transition. In this context, then-presidential candidate Gabriel Boric echoed these discussions and incorporated and prioritised the Just Socioecological Transition as one of the four pillars of his government programme (Gabriel Boric Electoral Campaign Team, 2021), thus bringing the Just Socioecological Transition into another sphere of social and political life (Fig. 5).

The global narrative of the Just Transition is understood as one with historical and political content, which, like many projects, is difficult to

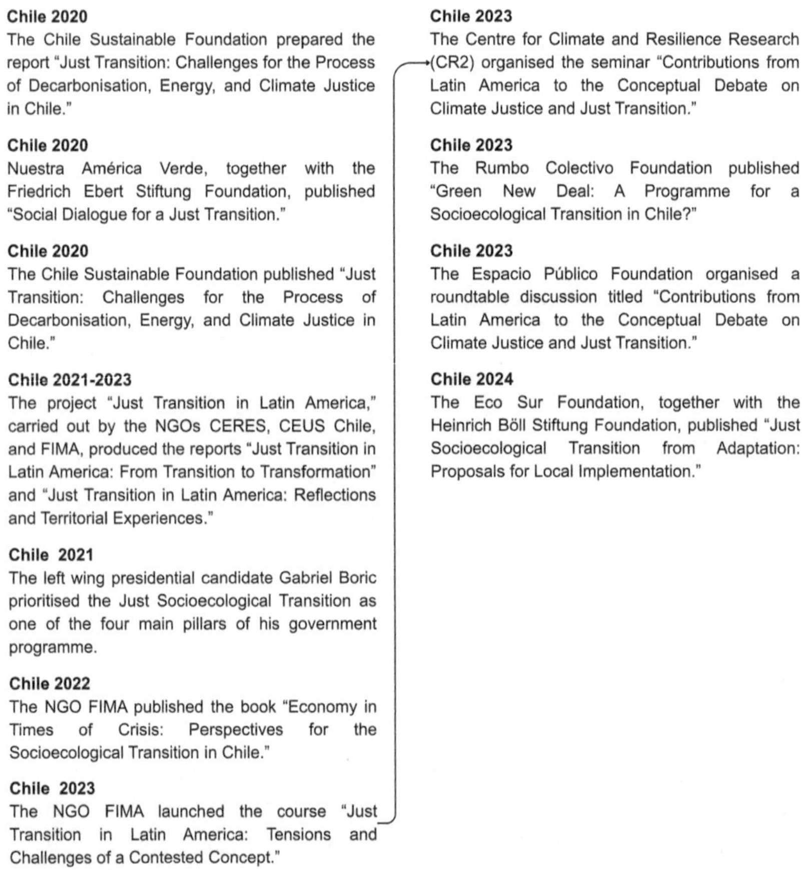

Fig. 5 Timeline showing the emergence of ethical–political debates and civil society reports on the Just Transition in Chile and Latin America. *CERES* Coalition for Environmentally Responsible Economies, *FIMA* Environmental Prosecutor's Office, *NGO* nongovernmental organisation

implement in Chile. The characteristics of an economy based on natural resource extraction, together with a post-dictatorship neoliberal social structure that undermined the transformative capacity of labour unions, make Chile a particular case. In this sense, one of the major advances of this genesis stage is the revival of the ethical debate on justice in the context of the Latin American transition, where environmental movements could become actors mobilising the ideals of justice.

3.3 *The State's Stabilising Experiment: The Just Socioecological Transition as an Object of Public Management*

With Gabriel Boric's inauguration as president in March 2022, a new phase began for the Just Socioecological Transition. Along with the appointment of renowned scientist María Heloísa Rojas as Minister of the Environment, an implementation process was launched (see Fig. 6), including the initiation of a legislative debate to regulate the Just Socioecological Transition process (Chilean Chamber of Deputies, 2022). Thus, a proposal was announced to amend the general environmental principles law. In addition, an Office for the Just Socioecological Transition was created within the Ministry of the Environment, tasked with coordinating various plans, measures, and policies in this direction that required a coordinating institution, such as the PRAS (Ministry of the Environment, Government of Chile, 2022b).

In April 2023, the Ministry of the Environment convened other ministries and created an Interministerial Committee for the Just Socioecological Transition (Ministry of the Environment, Government of Chile, 2022a). This established an institutional space for convergence and coordination between the Ministries of the Environment, Economy, Energy, Labour, Health, and Social Development, as well as collaboration with the Ministries of the Interior, Mining, Women and Gender Equity, and Education. This space is significant for defining a political strategy for implementing this ethical–theoretical concept within Chilean public administration.

The Just Socioecological Transition has become a predominantly state-driven narrative, and, as such, it is also technified and operationalised by public policy managers and decision-makers. Consequently, there has also been a process of closure, of stabilisation regarding what will—and will not—be considered a measure of the Just Socioecological Transition. For example, an institutional definition is provided: "Just

Chile 2022
Gabriel Boric assumed the presidency of Chile and appointed the climatologist María Heloísa Rojas Corradi as Minister of the Environment.

Chile 2022
Máximo Pacheco assumed the presidency of the National Copper Corporation of Chile (CODELCO), with the intention of closing highly polluting and economically inefficient operations.

Chile 2022
The legislative debate process began for Bill 15172–12, which regulates the process of a Just Socioecological Transition towards carbon neutrality and amends Law 19,300 on the General Bases of the Environment.

Chile 2022
The Ministry of the Environment created the Office for the Just Socioecological Transition to coordinate efforts in the so-called transition zones. This office manages the Environmental and Social Remediation Programmes (PRAS) in Huasco, Quintero, and Coronel, and supports territories with scheduled coal power plant closures, incorporating Tocopilla and Mejillones. In addition, it supports areas affected by polymetallic pollution in Arica and Calama, effectively overseeing seven transition territories.

Chile 2022
The Interministerial Committee for the Just Socioecological Transition was created as a coordination mechanism for the planned closures of coal-fired power plants and polluting industries. At the same time, it sought to ensure coordination and commitment to PRAS measures, which, although previously defined, lacked economic and political support from ministries. The committee was permanently composed of the Ministry of the Environment, the Ministry of Economy, Development and Tourism, the Ministry of Energy, the Ministry of Labour and Social Security, the Ministry of Health, and the Ministry of Social Development and Family, and also included collaboration from the Ministry of the Interior, the Ministry of Mining, the Ministry of Women and Gender Equity, and the Ministry of Education.

Chile 2023
The Government of Chile and CODELCO announced the closure of the Ventanas Smelter, a highly polluting and economically inefficient facility.

Chile 2023
The Citizen Fair for the Just Socioecological Transition Plan was held in Tocopilla.

Chile 2023
The Just Socioecological Transition project advanced in Congress and was approved by the Chamber of Deputies.

Chile 2023
The Tribu and Chile Foundations published the document "Productive Reconversion Strategy for Quintero–Puchuncaví: Strategic Roadmap."

Fig. 6 Two-part timeline showing the institutional consolidation and territorial governance of the Just Socioecological Transition in Chile from March 2022 to July 2024

Socioecological Transition is a concept we have coined for a series of transitions that Chile must undergo in order to move from a situation of environmental degradation, climate crisis, and biodiversity loss (affecting the population), towards a model that provides greater well-being for Chilean men and women" (Ministry of the Environment, Government of Chile, 2023).

Chile 2023
The Ministry of Energy published the Just Socioecological Transition Plan for Tocopilla.

Chile 2024
The Ministry of Energy submitted the Just Socioecological Transition Plan for Tocopilla to the Ministry of the Environment.

Chile 2024
Fundación Tribu published the document "Just Socioecological Transition of Coronel (JSJ-Coronel): Attract, Expand, and Create."

Chile 2024
Fundación Tribu published the document "Productive Reconversion Strategy for Tocopilla: Roadmap 2024–2050."

Chile 2024
The Ministry of the Environment focused the 2024 Environmental Protection Fund on financing citizen initiatives for the Just Socioecological Transition in Huasco, Coronel, Quintero, and Puchuncaví.

Fig. 6 (Coninued)

3.4 *What Comes Next? Prospects for the Just Socioecological Transition*

The association of the Just Socioecological Transition with a left-wing government presents two challenges: first, to ensure that the Just Socioecological Transition becomes a state narrative, and second, to ensure its continuity over time, independent of the government in power.

Anticipating a possible change in government, the executive branch developed national–global mechanisms to support the survival of the Just Socioecological Transition (see Fig. 7). First, the UNDP was tasked with developing a national public policy instrument: the National Strategy for a Just Socioecological Transition. This process included citizen participation, integrating public, private, and civil society actors. The aim was to ensure that the instrument would have social legitimacy, beyond the actors who promoted the Just Socioecological Transition from its inception. Second, this proposal would be submitted to the Committee of Ministers for approval. Third, the president would sign a decree to formally establish the National Strategy for a Just Socioecological Transition.

Chile 2022
At COP27, Chile committed to the Just Socioecological Transition. This commitment made it possible to address the demands of other territories affected environmentally and in labour terms that had not been included within the management scope of the Office for the Just Socioecological Transition due to resource limitations. These include areas such as Til-Til, shellfish gatherers in the south, salmon farming regions, and Andacollo with its mining-related environmental liabilities, among others.

Chile 2023
Decree No. 30 was issued, formalising an agreement with UNDP and the United Nations to support the development of the National Strategy for the Just Socioecological Transition. The agreement was ratified in February 2024.

Chile 2023
The Chile Strategy for the Implementation of the 2030 Agenda was published, incorporating the Just Socioecological Transition.

Chile 2024
The Council of Ministers for Sustainability and Climate Change approved a key step in the creation of the National Strategy for the Just Socioecological Transition.

Chile 2024
The concept appeared as a reference, without being defined, in Decree No. 1, which approved the National Policy on Science, Technology, Knowledge, and Innovation.

Chile–Spain 2024
The Ministry of the Environment signed a cooperation agreement on Just Ecological Transition with Spain's Ministry for the Ecological Transition and the Demographic Challenge.

Chile 2024
An Expert Panel was formed to support the development of the National Strategy for the Just Socioecological Transition.

Chile 2024
The citizen participation process for the development of the preliminary draft (anteproyecto) of the National Strategy for the Just Socioecological Transition was launched.

Chile 2024
A digital platform was launched to support citizen participation in the preparation of the preliminary draft of the National Strategy for the Just Socioecological Transition.

Chile 2024
Resolution No. 5586 approved the preliminary draft of the National Strategy for the Just Socioecological Transition, developed with citizen participation, and submitted it for public consultation.

Chile 2024
The public consultation process began for the draft of the National Strategy for the Just Socioecological Transition.

Chile 2024
The Ministry of the Environment organised webinars and workshops in all regions to carry out public consultation processes and to develop the draft of the National Strategy for the Just Socioecological Transition.

Fig. 7 Two-part timeline displaying the development of the Just Socioecological Transition strategy in Chile between November 2022 and November 2024. *COP27* 27th Conference of the Parties to the United Nations Framework Convention on Climate Change, *UNDP* United Nations Development Programme

Finally, once approved, the Strategy would contribute to the social pillar of the nationally determined contribution to the Paris Agreement. This link was highly significant, as it sought to provide the strategy with normative and legal force through a national–global framework.

4 Is it the Same but not Identical? Reflections on Innovation and Technopolitical Assemblages of Justice in Transitions

Researchers in Science and Technology Studies (STS) over the past decade have developed a dynamic dialogue with political theory and public policy. These studies understand that social and political life is based on the integration between humans and the material world (Barry, 2012; Braun & Whatmore, 2010; Ureta, 2023; Williams et al., 2024). These perspectives have revived the notion of assemblage to understand public policy (Ureta, 2015, 2023). In contrast to the stable conception (McCann & Ward, 2012), they view public policies as distributed, extended, and not entirely stable entities, always in flux, incorporating or discarding content and objectives (Gill et al., 2017).

Considering this notion of assemblage, it is difficult to draw clear distinctions and connections between the Just Transition and the Just Socioecological Transition. However, we identify three points of connection: first, the recognition of transitions as temporally extended phenomena; second, the understanding of transitions as political phenomena; and third, their ethical–political concern for the most disadvantaged or openly excluded actors. At the same time, we also identify three points of divergence between both approaches to justice: first, their social and political trajectories; second, the scale at which they project their action; and third, the actors and objectives they seek to transform (Fig. 8). None of these aspects can be considered permanent or settled, as the inherent dynamism of both policy approaches may lead to further synchronisations and desynchronisation.

Regarding their common aspects, both approaches understand that transitions are not isolated, one-off events. On the contrary, they recognise that transitions are historically situated processes, carrying contextual histories and oriented towards the future. Both approaches also acknowledge that transitions are not merely technological processes but also political phenomena, imbued with ethical and normative notions about how society should function. Finally, the Just Transition and the Just Socioecological Transition share an ethical–political concern not to deepen pre-existing inequalities or to perpetuate processes of social, cultural, or economic exclusion and marginalisation.

Connections	Differences
Transition as a temporally extended phenomenon	Specific social and political trajectory
Concern for disadvantaged or excluded actors	Global scale of action
Transformations understood as a political phenomenon	Context of transformation

Fig. 8 Matrix comparing connections and differences between two justice-based transition approaches: the Just Transition and the Just Socioecological Transition

With respect to their differences, the Just Transition has a predominantly labour union origin in the Global North, which has evolved to incorporate environmental issues in the context of the climate crisis. In contrast, the Just Socioecological Transition emerges from civil society and environmental movements and is later integrated into public administration. There is also a distinction in the scale of action and implementation: the Just Transition has operated at multiple scales, including the international level, whereas the Just Socioecological Transition has been more anchored in the Chilean context. Finally, there are also some distinctions regarding the actors and transformation objectives: although the Just Transition today includes more content, it continues to be politicised by labour and union actors with a climate orientation. In contrast, the Just Socioecological Transition, at least in its phase linked to civil society, incorporates a broader vision of justice, expressing concern for ecological functioning.

5 Conclusions

In this chapter, we have analysed the construction and evolution of two justice-oriented transition narratives: the Just Transition and the Just Socioecological Transition. Through historical reconstruction and comparative analysis, we have identified both their points of convergence and their differences.

On the one hand, the Just Transition originated within the labour movement of the Global North and, over the decades, has expanded its scope to become a key discursive framework in the international climate debate. Its initial focus on worker protection has broadened into a heterogeneous vision that incorporates new dimensions, topics, and actors. In contrast, the Just Socioecological Transition emerged in Chile from civil society debates and was later taken up by the state, with a particular emphasis on the management of territories affected by extractivism, which are recognised as territories in transition.

Despite their differences, both approaches share three central elements: the recognition of transitions as historical and prolonged processes, the understanding of these processes as political phenomena involving power struggles, and the concern to prevent changes in productive models from reproducing pre-existing inequalities. However, they also present significant distinctions in terms of their origins, scales of action, and objectives.

From the perspective of technopolitical assemblages, this analysis allows us to understand that justice-oriented transition approaches are also dynamic configurations that emerge from interactions among actors, institutions, and discourses. In this sense, the Just Socioecological Transition is neither a simple transposition of the Just Transition to the Chilean context nor an entirely indigenous creation. Rather, the Just Socioecological Transition is an assemblage defined by local debates, political tensions, and processes of institutional stabilisation.

Future empirical work should also analyse in greater depth the tensions and critical perspectives regarding the process of technification of justice in transitions. The consolidation of the Just Socioecological Transition within the state apparatus can be seen as a substantive advance, but it must also be analysed as part of a broader process of institutional and environmental legislative transformation, which also includes setbacks—for example, the publication of the so-called *permisología* law in Chile (Ojeda-Pereira, 2025). In its process of assemblage with the state, the Just Socioecological Transition may also lose some of the content that made it distinctive and may fall into political domestication that requires more thorough analysis.

Acknowledgements This chapter was supported by the National Agency for Research and Development (ANID), through the National Doctoral Scholarship, grant number 21221269. The author also acknowledges Professor Antti Silvast (LUT University) for his careful reading and valuable comments on earlier versions of this chapter.

References

Avelino, F., Grin, J., Pel, B. & Jhagroe, S. (2016). The politics of sustainability transitions. *Journal of Environmental Policy & Planning, 18*(5), 567–567.

Avelino, F., Grin, J., Pel, B., & Jhagroe, S. (2021). The politics of sustainability transitions. *Journal of Environmental Policy & Planning, 18*(5), 501–518. https://doi.org/10.1080/1523908X.2016.1216782

Aykut, S. C., & Maertens, L. (2021). The climatization of global politics: Introduction to the special issue. *International Politics, 58*(4), 501–518. https://doi.org/10.1057/s41311-021-00325-0

Barry, A. (2012). *Material politics: Disputes along the pipeline*. Wiley-Blackwell.

Braun, B., & Whatmore, S. J. (2010). *Political matter: Technoscience, democracy, and public life*. University of Minnesota Press. https://www.jstor.org/stable/10.5749/j.ctttv6sg

Campos-Medina, F., Korneeva, E., Krayneva, R., & Ojeda-Pereira, I. (2021). *Challenge to sustainability; the territorial agglomeration of renewable energies projects in central-South Chile* (pp. 32–39). https://doi.org/10.2991/aebmr.k.211118.007

Campos-Medina, F., Ojeda Pereira, I., & Ponce Labra, P. (2021). Extractivism in Chile: Three keys to understanding the logic of agglomeration of industrial projects and proposing alternatives for resistance. In C. Alister, X. Cuadra, D. Julián-Vejar, P. Blaise, & O. Camila (Eds.), *Questioning the neoliberal extractivist model from the South. Capitalism, Territories, and Resistances* (p. 57–85). Ariadna. https://doi.org/10.26448/ae9789566095316.18

Cavalheiro, G. (2020). *Initiating the just transition in Serbia. Leaving no one behind in the coal intensive regions and communities in Serbia*. United Nations Development Programme. https://www.klimatskepromene.rs/wp-content/uploads/2021/03/08-03-2021_Report-on-a-Roadmap-for-Just-Transition-for-Serbia.pdf

Chang, M., Thellufsen, J. Z., Zakeri, B., Pickering, B., Pfenninger, S., Lund, H., & Østergaard, P. A. (2021). Trends in tools and approaches for modelling the energy transition. *Applied Energy, 290*, 116731. https://doi.org/10.1016/j.apenergy.2021.116731

Chilean Chamber of Deputies. (2022). *Regulates the process of just socioecological transition towards carbon neutrality, and modifies for this purpose Law No. 19,300, general environmental framework Law*. Chamber of Deputies of

Chile. https://www.camara.cl/verDOC.aspx?prmID=68519&prmTipo=FICHAPARLAMENTARIA&prmFICHATIPO=DIP&prmLOCAL=0

Ciplet, D. (2022). Transition coalitions: Toward a theory of transformative just transitions. *Environmental Sociology*, *8*(3), 315–330. https://doi.org/10.1080/23251042.2022.2031512

Fuentes, C., Larraín, S., & Poo, P. (2020). *Just transition. Challenges for the decarbonization process, energy justice, and climate justice in Chile*. Chile Sustentable. https://www.chilesustentable.net/wp-content/uploads/2020/12/Transicion-Justa-baja.pdf

Gabriel Boric Electoral Campaign Team. (2021). *Programmatic proposals for the new Chile*. https://www.alejandrobarros.com/wp-content/uploads/2021/08/Boric.pdf

Geels, F. W., & Schot, J. (2007). Typology of sociotechnical transition pathways. *Research Policy*, *36*(3), 399–417. https://doi.org/10.1016/j.respol.2007.01.003

Gill, N., Singleton, V., & Waterton, C. (2017). The politics of policy practices. *The Sociological Review*, *65*(2), 3–19. https://doi.org/10.1177/0081176917710429

Gollop, F. M., & Roberts, M. J. (1983). Environmental regulations and productivity growth: The case of fossil-fueled electric power generation. *Journal of Political Economy*, *91*(4), 654–674. https://www.jstor.org/stable/1831072

Gordon, R. (1998). 'Shell No!': OCAW and the labor-environmental alliance. *Environmental History*, *3*(4), 460–487. https://doi.org/10.2307/3985207

Government of Chile. (2020). *Chile's nationally determined contribution (NDC). 2020 update*. Government of Chile. https://cambioclimatico.mma.gob.cl/wp-content/uploads/2020/08/NDC_2020_Espanol_PDF_web.pdf

Groves, D., Molina-Perez, E., Syme, J., Alvarado, G., De León, F., Acuña, J., & Jenkins, A. (2022). *A green recovery from COVID-19 in Costa Rica. Aligning decarbonization investments in Costa Rica with economic recovery*. United Nations Development Programme. https://www.undp.org/es/costa-rica/publications/una-recuperacion-verde-del-covid-19-en-costa-rica-alineando-las-inversiones-en-descarbonizacion-de-costa-rica-con-la

Hampton, P. (2015). *Workers and trade unions for climate solidarity: Tackling climate change in a neoliberal world*. Routledge.

ICFTU [International Confederation of Free Trade Unions]. (1997). *Climate change and jobs: Towards a strategy for sustainable employment. Trade union statement to the Kyoto conference*. The Global Development Research Centre. https://www.gdrc.org/ngo/azza/english/Part-I.pdf

Jenkins, K., Sovacool, B., Błachowicz, A., & Lauer, A. (2020). Politicising the just transition: Linking global climate policy, nationally determined contributions and targeted research agendas. *Geoforum*, *115*, 138–142. https://doi.org/10.1016/j.geoforum.2020.05.012

Lecourt, J., Fontecilla, F., & Pino, F. (2023). *Just transition in Latin America: Reflections and territorial experiences.* Just transition in Latin America project (TJLA), coordinated by CEUS Chile, ONG FIMA, Centro de Análisis Socioambiental.

McCann, E., & Ward, K. (2012). Assembling urbanism: Following policies and 'studying through' the sites and situations of policy making. *Environment and Planning A: Economy and Space*, *44*(1), 42–51. https://doi.org/10.1068/a44178

Ministry of Energy, Government of Chile. (2021). *Just transition strategy in the energy sector.* Government of Chile. https://energia.gob.cl/sites/default/files/documentos/estrategia_transicion_justa_2021.pdf

Ministry of the Environment, Government of Chile. (2022a). *Decree 57. Creates the interministerial committee for a just socioecological transition.* National Congress Library. https://www.bcn.cl/leychile/navegar?idNorma=1191447

Ministry of the Environment, Government of Chile. (2022b). *Office for a just socioecological transition.* https://mma.gob.cl/transicion-socioecologica-justa/

Ministry of the Environment, Government of Chile. (2023). *Just socioecological transition.* https://mma.gob.cl/contacto/preguntas-frecuentes/transicion-socioecologica-justa/

Morena, E., Krause, D., & Stevis, D. (Eds). (2020). *Just transitions: Social justice in the shift towards a low-carbon world.* Pluto Press. https://doi.org/10.2307/j.ctvs09qrx

Morena, E., Stevis, D., Shelton, R., Krause, D., Mertins-Kirkwood, H., Price, V., Azzi, D., & Helmerich, N. (2018). *Mapping just transition(s) to a low-carbon world.* UNRISD. https://www.unrisd.org/en/library/publications/mapping-just-transitions-to-a-low-carbon-world

Ojeda-Pereira, I. (2025). Chile's framework law erodes institutions. *Science*, *390*(6770), 249–249. https://doi.org/10.1126/science.aea4024

Ojeda-Pereira, I., Campos-Medina, F., & Gajardo, N. (2024). Cartographies of justice in the era of transitions toward sustainability, global and Latin American assemblages. *Polis (Santiago)*, *23*(68), 10–46. https://doi.org/10.32735/S0718-6568/2024-N68-3684

Pesch, U. (2015). Tracing discursive space: Agency and change in sustainability transitions. *Technological Forecasting and Social Change*, *90*, 379–388. https://doi.org/10.1016/j.techfore.2014.05.009

Pinto, A. P., García, P. B., Almonacid, L. E., & Henríquez, B. J. (2023). Sacrifice zones and socio-environmental recovery in Chile: Failures and opportunities in environmental policy. *Ambiente & Sociedade*, *26*, e01371. https://doi.org/10.1590/1809-4422asoc0137r1vu2023L3AO

Rabi, V., Pino, F., Fontecilla, F., & Lecourt, J. (2021). Just transition in Latin America: From transition to transformation. *Just Transition in Latin America Project (TJLA), Coordinated by NGO CERES, CEUS Chile, and NGO FIMA.*

Stevis, D., & Felli, R. (2016). Green transitions, just transitions? Broadening and deepening justice. *Kurswechsel, 3*, 35–45. http://www.beigewum.at/wp-content/uploads/KuWe-3_16-Stevis-Felli_Green-Transition-Just-Transition.pdf

Stevis, D., Morena, E., & Krause, D. (2020). Introduction: The genealogy and contemporary politics of just transitions. In E. Morena, D. Krause, & D. Stevis (Eds.), *Just transitions: Social justice in the shift towards a low-carbon world* (pp. 1–31). Pluto Press. https://doi.org/10.2307/j.ctvs09qrx

Svenja, K., Andersen, T., Simas, M., & Harsdorff, M. (2021). *Measuring the socioeconomic impacts of climate policies to guide NDC enhancement and a just transition.* United Nations Development Programme & International Labour Organization. https://www.undp.org/zimbabwe/publications/measuring-socioeconomic-impacts-climate-policies-guide-ndc-enhancement-and-just-transition

Tironi, M. (2018). Hypo-interventions: Intimate activism in toxic environments. *Social Studies of Science, 48*(3), 438–455. https://doi.org/10.1177/0306312718784779

UNDP [United Nations Development Programme]. (2022). What is a just transition? And why is it important? UNDP climate promise. https://climatepromise.undp.org/es/news-and-stories/que-es-la-transicion-justa-y-por-que-es-importante

United Nations Global Compact. (2022). *Just transition for climate adaptation: A business brief.* UN Global Compact. https://unglobalcompact.org/library/6099

Ureta, S. (2015). *Assembling policy: Transantiago, human devices, and the dream of a world-class society.* MIT Press. https://doi.org/10.7551/mitpress/9654.001.0001

Ureta, S. (2023). *Political experiments. Rethinking the implementation of public policies.* Universidad Alberto Hurtado.

Williams, R., Silvast, A., & Musiani, F. (2024). Governance by information infrastructures: Origins and evolution of the concept. *First Monday, 29*(10). https://doi.org/10.5210/fm.v29i10.13794

Working Lives Research Institute. (2008). *A green and fair future. for a just transition to a low carbon economy.* Trade Union Congress. https://www.tuc.org.uk/sites/default/files/documents/greenfuture.pdf

CHAPTER 3

The Ecological Transition and the Green Deal: A Review of the (Counter)currents Toward A Just Transition

José Miguel Fuentes-Zuleta

1 Introduction

The climate crisis has compelled countries to implement adaptation and mitigation measures. Although there are various options such as reforestation, disaster risk management, and the revalorization of local and Indigenous knowledge, among others, the number of alternatives varies depending on the productive sector (Hickel et al., 2021).

To achieve this green economy, according to Pollin (2020), it is necessary to expand investment in industrial policies that promote energy efficiency and renewable energy. This requires fostering technical innovations and adapting existing clean energy technologies to the specific context of each country, combining industrial policy instruments with research, development, and preferential tax incentives for investment in clean energy (Banerjee et al., 2012).

J. M. Fuentes-Zuleta (✉)
Faculty of Architecture, Design and Urban Studies, Pontifical Catholic University of Chile, Santiago, Chile; Nucleo Patrimonios – NupatS, Santiago, Chile
e-mail: jmfuentesz@ucl.cl

E. Costa Cordella (ed.), *The Ecological Transition as a Challenge to Law*, Just Transitions,
https://doi.org/10.1007/978-3-032-22055-4_3

In this context, the Green New Deal emerges as a program aimed at fulfilling the objective of the 2016 Paris Agreement regarding the mitigation of greenhouse gas (GHG) emissions to prevent the average global temperature from rising by more than 1.5 °C.

However, Hickel and Kallis (2020) argue that despite the consensus on investment in green technologies, these have not been sufficient to achieve the goal of staying below 1.5 °C, suggesting that such policies are incompatible with economic growth. Authors such as Goya (2014) question the synonymy between industrial policy and economic growth as a path to development, noting that in Chile, growth has been associated with greater inequality and environmental degradation.

In response to this problem, it is argued that the ecological transition is being instrumentalized by green capitalism (Williams, 2024), hindering a just transition. This model does not address the structural causes of the environmental crisis, maintaining the capitalist rationality of consumption, profit, and technological optimism. Wallis (2010) argues that capitalism inherently exploits natural resources, making environmental collapse a systemic problem rather than a side effect. Therefore, attempts to solve the crisis within capitalism are limited.

Following the discussions raised by the Anthropocene, the current crisis does not stem from natural factors but is the result of human activity and the modern rationality underpinning the economic model (Chernilo, 2017) and the policies that have led to this environmental crisis (Tironi & Undurraga, 2023).

On this basis, this chapter will analyze how countercurrents that challenge green capitalism and institutional programs such as the Green Deal (specifically degrowth, post-growth, ecofeminism, and ecosocialism) expose their limitations, and propose alternative approaches to achieve a just transition. The hypothesis is that critiques of green capitalism have enabled the emergence of various countercurrents or critical discourses that question international programs—more or less consensual—such as the (New) Green Deals (both American and European), highlighting the limitations of these programs while proposing alternative possible futures.

It is concluded that these countercurrents, while critical of the principles of green capitalism, are not opposed to ecological transition. Each of these critical perspectives raises specific points of contention with the Green New Deal, with some adopting more counterhegemonic discourses than others. In this sense, this chapter will analyze and discuss degrowth, post-growth, ecofeminism, and ecosocialism.

2 The Green New Deal as a Response to the Climate Crisis

The Green Deals can be divided into two models proposed by the Global North: the American model (Green New Deal 1.0 and 2.0) and the European model (European Green Deal). Although each has its specificities, the common points between these three programs, which define the essence of the Green Deal, are the redesigns of environmental, fiscal, and industrial policies (Bloomfield & Steward, 2020).

In concrete terms, these Green Deals seek a profound transformation of the economic system along with measures to reduce socioeconomic inequality through massive public investment. The central (and shared) demands of both Green Deals are (1) transition to a coal-free society through a drastic reduction of GHG emissions, infrastructure renewal, and a commitment to energy efficiency; (2) reform of the national and international financial and tax sectors; and (3) significant progress in reducing socioeconomic inequality during the transition.

The following subsections review each of these Green Deal models and their respective characteristics and proposals for ecological transition.

2.1 *European Green Deal*

On December 11, 2019, the European Commission launched its proposal for a European Green Deal. This pact emerged in response to pressure from youth movements, Green parties in the European Parliament, and the scientific community (Giegold & Mack, 2012). Specifically, the European Green Deal aims to transform five key sociotechnical systems: energy, industry, construction, mobility, and food, which are responsible for the majority of carbon emissions (Bloomfield & Steward, 2020).

Rather than representing a green revolution, the European Green Deal focuses on the need to change systems and offer sustainable transition alternatives in the areas most harmful to the environment (Bloomfield & Steward, 2020). To achieve this transition, the European Green Deal proposes that public policies and politics must be transformed, reorienting macroeconomic processes in line with sustainable growth and establishing a relationship between environmental protection policies and sustainable development policies (Bloomfield & Steward, 2020).

The European Green Deal's strategies include investment in technology to both reduce CO_2 emissions and increase competitiveness by lowering production costs. In terms of policies and programs, the former involve combining national and European Union (EU) funds to accelerate resource efficiency, while the latter should be developed at the European and regional levels in a threefold partnership between private actors, the public sector, and academia (Bloomfield & Steward, 2020).

2.2 *North American Green New Deal*

The Green New Deal, or Green New Deal 1.0, emerged in the context of the 2008 financial crisis when a group of British economists created what would later be called the Green New Deal Group (GND Group). This initial Green New Deal focused on a set of public measures aimed solely at stimulating the economy through investment and facilitation of the energy transition.

Subsequently, in 2018, under the auspices of the Democratic Party and driven primarily by Alexandria Ocasio-Cortez, what is known as the Green New Deal 2.0 emerged. This version, however, goes beyond the proposal for ecological modernization. As Giegold and Mack argue, the Green New Deal 2.0 aims for a social and ecological transformation of the economy to achieve net zero GHG emissions (Giegold & Mack, 2012) through an energy revolution (Bauhardt, 2014) and the decarbonization of various sectors to the extent that is technologically feasible (Hickel et al., 2021).

The Green New Deal 2.0 calls for economic intervention to achieve decarbonization, emphasizing public investment, industrial policies, and indicative planning, while also highlighting the importance of activists and environmental justice movements. With regard to social transformation, the Green New Deal 2.0 proposes that coal-based production sectors should be completely replaced by renewable energy, which would create new jobs in emerging sustainable industries (Bauhardt, 2014). In this sense, the Green New Deal 2.0 advocates for a just transition, stating that "socially vulnerable groups should not be the losers in the ecological transformation" (Giegold & Mack, 2012).

3 Green New Deal: Green Capitalism or Structural Solutions?

All of these proposals put forward by the Green New Deal, both European and North American, fall within what Gramkow (2020) calls the big push for sustainability, referring to "a set of complementary sustainable investments that generate a virtuous circle of economic growth, job creation, development of productive chains, and reduction of environmental impacts, while restoring the productive capacity of natural capital" (Gramkow, 2020, p. 189). In this regard, Schpelman (2009) argues that a Green Deal should be more than a technological platform to support eco-industries, since even green economies can be harmful. Following this argument, the Green New Deal, according to They brazilian sociologist Thea Riofrancos (2019), presents certain problems, particularly with its goal of an energy transition focused on renewables. According to the Riofranco, this can be seen in two specific cases. The first issue relates to electromobility. Lithium plays a fundamental role in the transition to electric vehicles, as it is essential for the batteries that power these vehicles. To achieve this shift to zero-carbon mobility, northern Chile—located in the lithium triangle shared with Argentina and Bolivia, where lithium is extracted—will bear the cost, affecting the Indigenous communities living there (Aparicio, 2023). The second issue concerns the use of renewable energy, specifically wind and solar power, because of the large amount of space required for these installations. In this sense, as Giegold and Mack point out, a new industrial policy under the Green Deal "can only succeed if it reduces dependence on imports of fossil fuels and other nonrenewable raw materials" (2012, p. 5). According to Riofrancos (2019), this leads to a divergence between what is technologically feasible and what is ecologically feasible. As Bauhardt (2014) notes, the Green New Deal takes the current conditions of industrial capitalism as its starting point, does not question the basic functioning of capitalist production and reproduction, and, moreover, is not inherently critical of growth; it simply demands a different kind of growth. In this sense, the transition from the logic of green capitalism would be more of a greenwashing, proposing green initiatives that are largely superficial and do not address the underlying problem (Williams, 2024).

4 Countercurrents to the Green New Deal

In response to criticisms of the Green New Deal—which question its usefulness for an ecological transition, as it does not address the structural causes of the climate crisis—several currents have emerged that challenge the Green New Deal by questioning either green capitalism or certain elements proposed by the Green New Deal. In they following subsections, we describe the currents that are critical of the Green New Deal, which propose alternative visions for possible futures.

4.1 Degrowth

The main argument of the theory of limits to growth is that economic growth cannot continue indefinitely on a finite planet, leading to conflicts with ecological systems and socioenvironmental problems. According to Gómez-Baggethun, from the tradition of limits to growth, it is argued that solutions such as renewables replacing fossil fuels require significant investment in technology, as well as materials that are sometimes finite, such as lithium. As a result, the extractivist logic persists, and these solutions do not limit growth; rather, they enable continued growth in a more sustainable way, under the logic of green capitalism (Gómez-Baggethun, 2020). The concept of degrowth (or planned economic degrowth) arises from the contributions of two authors: Georgescu-Roegen (1995), who argued that degrowth could be a form of long-term survival for humanity, achievable as long as the physical, demographic, and cultural conditions exist; and Latouche (2010), who advocates for a shift away from production—and consumption-oriented growth, instead fostering social and relational wealth, such as cooperation, well-being, humanistic values, the health of nature, and local self-sufficiency. Both Georgescu-Roegen's and Latouche's theories include political and global dimensions, viewing the exploitation of natural resources in relation to environmental justice and social equality, rather than solely from an ecological perspective. Furthermore, degrowth seeks to replace indicators such as gross domestic product (GDP) with others that focus on the well-being of people. Proponents of sustainable development argue that economic degrowth provides the conditions under which environmental protection can best be achieved.

As noted by Hickel et al. (2021), the Green New Deal and degrowth converge on the importance of public investment to finance the energy transition, industrial policies to lead the decarbonization of the economy, socialization of the energy sector to enable greater investment horizons, and expansion of the welfare state to increase social protection.

In concrete terms, the Green New Deal advocates investing in renewable energy, which will expand certain activities and stimulate the economy. Economic growth, in turn, will increase the funds available for investment in clean energy. However, degrowth emerges as a direct response to the Green New Deal's reliance on economic growth. This school of thought proposes a planned reduction in consumption in the wealthiest economies, while ensuring an equitable redistribution of resources. Moreover, "degrowth is not only about government policies, but also about changes in values and ways of life" (Mastini, Kallis, and Hickel, p. 5).

Degrowth maintains that the Green New Deal does not sufficiently challenge the core principles of capitalism, such as consumerism or the logic of capital accumulation, and that without a change in these pillars, the transition to renewable energy would remain unsustainable. Furthermore, degrowth points out that some measures proposed by the Green New Deal can be implemented without economic growth, such as reduction of working hours, redistribution of wealth, green tax reform, and establishment of environmental limits.

However, unlike the Green New Deal, the set of ideas encompassed by degrowth is more closely associated with academic and activist circles, and it has yet to become a priority for political parties seeking to implement it. In the same vein, it is important to note that, unlike the Green New Deal, the success of degrowth necessarily depends on a cultural transformation.

4.2 Post-Growth

Post-growth holds that economic growth, specifically as measured by GDP, is not an adequate indicator of well-being, instead proposing quality of life, social equity, and ecological sustainability as reference points (Hickel et al., 2021), thus distancing itself from the Green New Deal. Rather than advocating for a green economy, post-growth proposes a restructuring of the economy toward a more local and competitive model (Bauwens, 2021).

For post-growth theorists, there are four priority areas to achieve a different kind of growth: new spatial patterns of production and consumption; regional and local transition strategies; adoption of more environmentally friendly and efficient modes of design, production, and resource use; and investment in renewable energy (Schulz & Bailey, 2014).

As Schulz and Bailey (2014) point out, post-growth should not be considered synonymous with contraction or recession; rather, it should be viewed as a concept like degrowth, which rejects the paradigm that prosperity can only be guaranteed through continuous material expansion and a monetarily measurable economy. As Hickel et al. (2021) note, for high-income countries, continuous economic growth may not be necessary. Instead, these countries can adopt post-growth policies that deliver improved social well-being without economic growth, without having to rely so heavily on negative emissions technologies or productivity gains.

Post-growth advocates maximizing the well-being of both humans and nonhuman life not only through job creation but also by building and empowering communities (Bauwens, 2021). Thus, the challenge of post-growth is not to pursue ever more desperate policies to recover the lost foundations of hyper-productivity driven by fossil fuels, but, rather, to create the conditions for an economy that works for everyone, within the limits of a finite planet (Jackson, 2019, p. 245).

Nevertheless, as Bauhardt (2014) observes, while post-growth challenges current patterns of production and consumption, it does not question the principles of the market economy. Post-growth, as Blühdorn (2017) notes—along with other "post-" concepts such as post-capitalism or post-consumerism—is less a model of policy and governance for (un) sustainability and more a discourse of simulation that does not entirely abandon the logic of growth; rather, it is directed at those countries that have already achieved full development and no longer need to continue growing.

4.3 Ecofeminism

Ecofeminism advocates for a care-based economy in which both reproductive and productive tasks are valued, calling for a profound restructuring of society that incorporates values such as solidarity, interdependence, and respect for ecological limits (Herrero, 2015). Ecofeminism thus criticizes the Green New Deal for failing to address the intersection of ecological crisis, patriarchy, and capitalism, arguing that the "ecofeminist political economy sees a connection between the exploitation of women's labour and the abuse of planetary resources" (Mellor, 2005, p. 123).

According to Kesting (2011), gender is a fundamental part of green economies. Ecofeminism provides an analysis of the current destructive relationship between humans and the nonhuman world by understanding the position of women at the margins of economic systems (Mellor, 2005). From this perspective, women—or, more precisely, their labor and lives—are, like the natural world, externalized and exploited for their economic value.

Ecofeminism finds support for its position in evidence that women tend to suffer more than men from the effects of environmental degradation and often take on greater responsibilities for ecosystem resilience (Kesting, 2011). Thus, ecofeminism criticizes the paradox of idealizing yet underestimating the productivity of care work and ecosystem services, a paradox that Bigo (2010) calls the "care crisis" or "care deficit."

One critique that ecofeminism makes of development ideology is that by prioritizing commodity production and economic growth, it destroys healthy and sustainable ways of life, leading to real material poverty, such as the loss of essential natural resources like water, fertile land, and biodiversity (Shiva, 2004).

According to Bauhardt (2014), ecofeminism offers alternatives to the current mode of production and consumption because it considers unpaid labor and human needs. The ecofeminist economic approach links the ecological crisis with the crisis of social reproduction, thus drawing a parallel between the exploitation of natural resources and the exploitation of women's labor, as both are necessary prerequisites for capitalism, regarded as free or costless elements that are naturally given, forming the basis for the growth of any market economy (Floro, 2012). This is why the ecofeminist political economy understands the ecological crisis and the crisis of social reproduction and care as two intrinsically intertwined processes (Bauhardt, 2014).

4.4 *Ecosocialism*

Ecosocialism posits a triple crisis within the capitalist model: economic, social, and ecological (Löwy, 2005). This school of thought is based on a critique of capitalism as an economic system, arguing that it is inherently unsustainable because of its constant need for growth, accumulation, and exploitation of natural resources (Schwartzman, 2009). This model questions green capitalism, maintaining that the attempt to resolve the crisis through technological solutions does not address the structural causes of the ecological crisis.

For Löwy (2005), socialism and environmentalism are intrinsically linked, as both share goals that involve questioning the current economic model and quantification as a form of value. According to Löwy (2005), both socialism and ecology appeal to qualitative values: for socialists, use value, satisfaction of needs, social equality; for environmentalists, protection of nature and ecological balance. Moreover, both conceive of the economy as being embedded within the environment, whether social or natural.

Following this critique of the economic model, solutions such as taxes or the green economy create the illusion of cleaner capitalism but do not change its ethos as such (Williams, 2024). Thus, Löwy (2005) argues, the problem with the proposals put forward by leading circles such as the EU is that the solutions offered for the ecological crisis are inadequate and inappropriate, as they do not recognize the connection between productivism and capitalism.

According to Löwy, the reasoning of ecosocialism rests on two essential arguments or principles. First, the current mode of production and consumption in advanced capitalist countries—based on the logic of unlimited accumulation (of capital, profits, and commodities), resource waste, conspicuous consumption, and the accelerated destruction of the environment—cannot possibly be extended to the entire planet without a major ecological crisis. Second, the continuation of capitalist progress and the expansion of a civilization based on a market economy will directly and inevitably threaten the very survival of the human species. Therefore, the protection of the natural environment is not only an ecological issue but also a humanitarian one. Achieving this would require a change not only in production but also in consumption.

According to Schwarzman (2009), the essential features of an ecosocialist transition include (1) application of the principles of containment and precaution to the environment (including industrial ecology and organic agriculture focused on and in green cities); (2) progressive dematerialization of technology and global availability of cutting-edge technology; and (3) increased human population density centered in green cities with the elimination of urban sprawl, leaving extensive biosphere reserves managed to preserve biodiversity.

In summary, ecosocialism encompasses those theories and movements that seek to subordinate exchange value to use value, organizing production as a function of social needs and the requirements of environmental protection. Its goal, ecological socialism, would be an ecologically rational society founded on democratic principles such as social equality and collective ownership of the means of production.

5 Critical Analysis of the Schools of Thought

Table 1, based on the aforementioned schools of thought, summarizes the four currents and the Green New Deal, comparing their views on growth, their proposals, the public policies they advocate, the strategies to achieve these, and the actors involved (or expected to be involved) in each.

The comparison between the different schools of thought reveals substantive differences regarding the role of economic growth and the forms of social transformation required in the face of the ecological crisis. While the Green New Deal maintains a position compatible with the idea of growth, degrowth proposals explicitly reject it, arguing that sustainability requires abandoning that logic entirely. Positioned between these stances is post-growth, which adopts an "agnostic" approach in that it does not consider growth either as a goal or as a problem; instead, it focuses on redesigning institutions to make a dignified life possible within planetary boundaries.

Both ecosocialism and ecofeminism share a structural critique of the current economic and political system, although they do so from different perspectives. Ecosocialism emphasizes the need to overcome capitalism in order to resolve the socioecological crisis, while ecofeminism introduces a unique perspective by linking gender relations, care, and historical inequalities to environmental degradation.

Regarding transformation strategies, the Green New Deal advocates for reform from within the system, seeking to redirect investments and public policies to reconcile well-being and sustainability. In contrast, both degrowth and ecosocialism propose deeper and more radical changes, which involve a comprehensive revision of the foundations of the contemporary economic model. Once again, post-growth occupies an intermediate position, adopting a pragmatic approach focused more on institutional design than on ideological rupture.

Finally, the actors supporting each of these proposals also differ. Some are backed by more institutional and governmental sectors, while others are sustained by social movements and collectives that drive more radical transformations.

Table 1 Comparison of the currents of degrowth, post-growth, ecofeminism, and ecosocialism in relation to the Green New Deal

	Green New Deal	*Degrowth*	*Post-growth*	*Ecofeminism*	*Ecosocialism*
View of growth	Green and sustainable growth	Explicit rejection of economic growth	Indifference to economic growth	Critique of growth from the perspective of care and social reproduction	Critique of capitalist growth
	Compatible with economic growth	Planned reduction of production and consumption	Priority given to social well-being	Critique of productivism and extractivism	Democratic planning of production
Proposal	Decarbonization of the economy	Decarbonization of the economy	Designing economies that function well without the need for growth	Reorganization of the economy centered on sustaining life	Radical transformation of the economic system under democratic control and ecological criteria
	Investment in renewable energy	Investment in renewable energy	Social and technological innovation oriented toward well-being	Feminist economics	Social ownership of the means of production
	Efficiency improvements	Reduction of social and economic metabolism (energy and resource use)	New metrics of progress beyond GDP	Food sovereignty and solidarity economy	International cooperation for a just transition

(continued)

Table 1 (continued)

	Green New Deal	*Degrowth*	*Post-growth*	*Ecofeminism*	*Ecosocialism*
Public policies	Public investment in renewable energy	Green tax reforms	Reform of the financial and monetary system	Recognition of care work	Nationalization of strategic sectors
	Just transition	Environmental and consumption limits and prohibitions	Progressive tax reform	Work–life balance policies	Democratic planning
	Guaranteed green jobs	Limits on resource consumption	Guaranteed employment policies	Gender-responsive budgeting	Worker and community control
	Strict environmental regulations	Universal basic services guarantees	Universal basic services guarantees	Protection of territories	Protection of the commons
	Carbon taxes	Reduction of working hours	Universal social provision of basic needs	Recognition of reproductive labor	Ecological agrarian reform
Strategies	Just energy transition	Changes in economic policies	Development of alternative indicators beyond GDP	Mutual aid networks	Social and political mobilization
	Infrastructure modernization	Decommodification	Strengthening public services	Defense of the commons	Worker self-management
	Changes in public investment (public–private financing)	Democratic negotiation of legitimate needs	Redistribution of wealth	Solidarity economy	International alliances

(continued)

Table 1 (continued)

	Green New Deal	*Degrowth*	*Post-growth*	*Ecofeminism*	*Ecosocialism*
	Grassroots activism	Voluntary simplification/self-limitation	Social innovation	–	Political education
Supporting actors	Progressive politicians	Left-wing green politicians	Reformist academics	(Eco)feminist movements	Radical left parties
	Environmental and social justice activists	(Radical) social movements	Development NGOs	Rural and Indigenous women's organizations	Anticapitalist movements
	Trade unions	Ecologist collectives	Social justice movements	Solidarity economy networks	Militant trade unions
	Moderate environmental NGOs	Eco-communities	Moderate progressive sectors	–	Indigenous movements
	Green businesses	–	–	–	–

GDP gross domestic product, *NGO* nongovernmental organization
Source Own elaboration

In turn, each of the currents criticizes the others in one way or another. These critiques reveal fundamental tensions regarding the role of economic growth, the degree of radical change required, political feasibility, and the integration of different perspectives (class, gender, and ecology). The climate justice movement, which is a recurring actor in degrowth and ecosocialism, criticizes the Green New Deal for failing to address global inequalities and the power asymmetries that structure the climate crisis.

Proponents of the Green New Deal, such as Pollin (2018), argue that growth is necessary to finance the transition to a green economy, contending that higher GDP levels will mean greater investment channeled into clean energy projects. Meanwhile, critics of growth converge in asserting the need for a fundamental socioecological transformation in industrialized societies that transcends the imperative of the growth paradigm (D'Alisa et al., 2014).

Critics of the Green New Deal argue that by operating primarily within the framework of developed countries, it fails to consider the disproportionate impact that climate change has on vulnerable communities in the Global South, nor does it acknowledge the impact that extractive activities required to generate low-carbon energy would have on these communities (Schmelzer, 2016).

Nevertheless, this "North–globalist" view is also applied to both degrowth and post-growth, as they are considered ahistorical in the sense that they ignore and underestimate long-term historical roots, path dependencies, and the power relations embedded in statistical standards (Schmelzer, 2016). Moreover, there are recurring patterns in the critiques: the more radical models criticize the insufficiency of reformist approaches, the more reformist models criticize the lack of political feasibility of the radical ones, and each model tends to view the integration of its perspective in the others as inadequate. Table 2 presents the cross-critiques.

On the basis of the analysis conducted, both from the comparison of the different currents and from the cross-critiques among them, it is possible to identify some points of convergence. While there are several differences—mostly related to a critique of economic growth and the prevailing economic model, capitalism—there are also certain points of convergence regarding shared objectives and challenges. The points of agreement and disagreement for each current in relation to the Green New Deal are presented in Table 3.

Table 2 Cross-critiques between the currents of degrowth, post-growth, ecofeminism, and ecosocialism in relation to the Green New Deal

	Green New Deal	*Degrowth*	*Post-growth*	*Ecofeminism*	*Ecosocialism*
Green New Deal	–	Maintains the logic of infinite economic growth	Lacks a clear vision of how a steady-state economy would function	Does not integrate gender and care perspectives	Does not question private ownership of the means of production
		Relies excessively on technological solutions	Does not sufficiently consider the limits to green growth	Maintains a masculine productivist logic	Maintains capitalist power relations
		Does not sufficiently question consumption patterns	Depends excessively on public debt	Ignores reproductive and care work	Is insufficient for a genuine systemic transformation
		May increase mineral extraction for green technologies	Does not address the contradictions of the current financial system	Does not question the sexual division of labor	Does not address ecological imperialism
Degrowth	Too radical to be politically viable	–	Excessive emphasis on reduction rather than transformation	Does not sufficiently integrate a feminist perspective	Insufficient emphasis on class struggle
	Could generate social and political resistance		May alienate sectors necessary for change	Neglects patriarchal power relations	Does not clearly articulate the political subject of change

(continued)

Table 2 (continued)

	Green New Deal	*Degrowth*	*Post-growth*	*Ecofeminism*	*Ecosocialism*
	Does not offer clear solutions for employment Difficult to implement in the Global South		Negative framing that hinders social acceptance Lack of concrete institutional proposals	Insufficient attention to the care economy Does not adequately consider reproductive labor	May be functional to capital in its reformist version Neglects the international dimension of struggle
Post-growth	Too agnostic regarding the role of growth Lack of clarity in concrete proposals	Insufficiently radical in its critique of growth Too reformist	–	Neglects the gender dimension Does not sufficiently integrate care	Does not sufficiently question property relations Lack of class analysis
	May discourage necessary investment	Does not sufficiently question consumerism		Maintains androcentric biases	Insufficient reformism
	Does not take advantage of the potential of green technology	May legitimize business as usual		Insufficient attention to social reproduction	Does not articulate a clear power strategy
Ecofeminism	Insufficient attention to technological solutions	Insufficient critique of economic growth	Lack of concrete institutional proposals	–	Insufficient class analysis

(continued)

Table 2 (continued)

	Green New Deal	*Degrowth*	*Post-growth*	*Ecofeminism*	*Ecosocialism*
	May be overly localist	May fall into gender essentialism	May be overly abstract		May dilute the capital–labor contradiction
	Lack of scalability of proposals	Lack of clear macroeconomic proposals	Difficulty scaling up alternatives		Risk of essentialism in views of nature
	Difficulty articulating concrete public policies	Excessive emphasis on the local rather than the systemic	Insufficient attention to systemic transformation		Lack of a strategy for taking power
Ecosocialism	Too radical to be politically viable	Maintains productivist biases	Excessively ideological	Reproduces androcentric biases of Marxism	–
	Excessive emphasis on centralized planning	Insufficient critique of industrialism	Lack of concrete institutional proposals	Insufficient attention to social reproduction	
	May discourage innovation	Excessive reliance on state planning	Excessive emphasis on seizing state power	Maintains the sexual division of labor	
	Risks of bureaucratization	Does not sufficiently question technological development	Neglects bottom-up social innovations	Does not adequately integrate care	

Source Own elaboration

Table 3 Points of agreement and disagreement between the currents of degrowth, post-growth, ecofeminism, and ecosocialism

	Degrowth	*Post-growth*	*Ecofeminism*	*Ecosocialism*
Points of agreement	Both recognize the need to transform the current model of production and consumption	Both seek alternatives to the current economic model and recognize the need for indicators beyond GDP	Both emphasize social justice and acknowledge the gendered dimensions of the climate crisis. They share a critique of the extractivist model	Both recognize the need for systemic transformation and the interconnection between environmental crises and social inequalities. They share a vision of a just transition and the importance of the public sector
Points of disagreement	Green New Deal maintains a vision of "green growth" and ecological modernization, while degrowth fundamentally questions the possibility of sustainable growth. Degrowth proposes a planned reduction of production and consumption	Post-growth is more radical in its questioning of growth as a social and economic objective. Green New Deal retains elements of growth-oriented logic in its transition proposal	Ecofeminism offers a deeper critique of the patriarchal system and its relationship to environmental destruction. It also more radically questions anthropocentrism, which is less present in Green New Deal	Green New Deal maintains a more reformist stance within the capitalist framework, while ecosocialism proposes a complete rupture with the capitalist system. Ecosocialism is more radical in its critique of economic growth

GDP gross domestic product
Source Own elaboration

6 Conclusions

As can be seen, despite differences regarding the discourse on growth, there are convergences and shared challenges concerning the urgency of addressing the climate crisis and the need to link environmental solutions with social justice. Table 4 groups these three points.

First, there are convergences regarding the need for systemic transformation, which generally involves decarbonization; a critique of the current economic model, whether through a Keynesian model such as the Green New Deal or a more radical change such as socialism; the importance of social justice, upholding the idea of a just transition; and the recognition of planetary boundaries as a diagnosis of the current situation.

Regarding shared challenges, all currents agree on the importance of building sufficient political support to back state policies; managing a just transition for workers; confronting resistance from privileged sectors such as large corporations; articulating viable alternatives, such as traditional knowledge or worldviews of Indigenous peoples; and coordinating actions at multiple scales.

Finally, regarding differences, there is the role of economic growth: while for the Green New Deal it is compatible with the ecological transition, for models such as ecosocialism or degrowth it is not; along these lines, they differ in the degree of rupture with capitalism, their strategies, and the temporal scale of transformation.

Table 4 Points of convergence, main differences, and shared challenges concerning the urgency of addressing the climate crisis and the need to link environmental solutions with social justice

Points of convergence	*Main differences*	*Shared challenges*
Need for systemic transformation	Role of economic growth	Building sufficient political support
Critique of the current economic model	Degree of rupture with capitalism	Managing a just transition
Importance of social justice	Transformation strategies	Facing resistance from privileged sectors
Recognition of planetary limits	Main agents of change	Articulating viable alternatives
	Temporal scale of transformation	Coordinating action at multiple scales

Source Own elaboration

In conclusion, the Green New Deal represents a more moderate and pragmatic position in comparison with the other currents, seeking changes within the existing system. In this sense, the Green New Deal could be seen as an entry point for deeper transformations, although for the other currents it is insufficient to address the crisis. The other currents share a more radical critique of the economic growth paradigm and, except for post-growth, of the capitalist system. It is worth noting that although each model has its differences, these should be observed and analyzed on the basis of the particularities of each state, considering their geopolitical position, whether from the Southern Cone or the Global North. Therefore, rather than choosing one over the others, it would be necessary to seek a new path where their convergences come together.

References

Aparicio, E. (2023). *Without Territory There Is No Culture: The Impact of Mining on Colla Communities*. Climate Tracker Latam.

Banerjee, R., et al. (2012). *Global Energy Assessment–Toward a Sustainable Future: Key Findings; Summary for Policymakers; Technical Summary*. Cambridge University Press.

Bauhardt, C. (2014). Solutions to the crisis? The Green New Deal, Degrowth, and the Solidarity Economy: Alternatives to the capitalist growth economy from an ecofeminist economics perspective. *Ecological Economics, 102*, 60–68.

Bauwens, T. (2021). Are the circular economy and economic growth compatible? A case for post-growth circularity. *Resources, Conservation and Recycling, 175*, 2.

Bigo, V. (2010). The care paradox: Devaluing and idealising care, the mother, and mother nature. *International Journal of Green Economics, 4*(2), 117–133.

Bloomfield, J., & Steward, F. (2020). The politics of the green new deal. *The Political Quarterly, 91*(4), 770–779.

Blühdorn, I. (2017). Post-capitalism, post-growth, post-consumerism? Eco-political hopes beyond sustainability. *Global Discourse*. doi: https://doi.org/10.1080/23269995.2017.1300415

Chernilo, D. (2017). The question of the human in the Anthropocene debate. *European Journal of Social Theory, 20*(1), 44–60.

D'Alisa, G., Demaria, F., & Kallis, G. (2014). *Degrowth: A vocabulary for a new era*. Routledge.

Floro, M. (2012). The crises of environment and social reproduction: Understanding their linkages. *Journal of Gender Studies, 15*, 13–31.

Georgescu-Roegen, N. (1995): La Décroissance. Entropie – Écologie – Économie. 2nd edn. Éditions Sang de la terre, Paris.

Giegold, S., & Mack, S. (2012). No stabilization of the €uro without a Green New Deal. Strategic paper by the Greens/EFA in the European Parliament. http://greennewdeal.eu/green-economy/publications/2011/en/no-stabilization-of-theeuro-without-a-green-new-deal.html

Gómez-Baggethun, E. (2020). More is more: Scaling political ecology within limits to growth. *Political Geography, 76*, 102095.

Goya, D. (2014). *Industrial policy: What it is, why it is necessary, and its past, present, and future in Chile*. Working paper, Network of Studies for the Deepening of Democracy (RED)

Gramkow, C. (2020). The Big Push for Sustainability: Shifting development models offers an opportunity to build resilient, low-carbon economies while tackling longstanding structural problems in Latin America and the Caribbean. *NACLA Report on the Americas, 52*(2), 186–191.

Herrero, Y. (2015). Introductory notes on ecofeminism. *Information Resources Bulletin, 43.*

Hickel, J., et al. (2021). Urgent need for post-growth climate mitigation scenarios. *Nature Energy, 6*(8), 766–768.

Jackson, T. (2019). The post-growth challenge: Secular stagnation, inequality and the limits to growth. *Ecological Economics, 156*, 236–246.

Kesting, S. (2011). What is 'green' in the green new deal—criteria from ecofeminist and post-Keynesian economics. *International Journal of Green Economics, 5*(1), 49–64.

Latouche, S. (2010). Degrowth. *Journal of Cleaner Production, 18*(6), 519–522.

Löwy, M. (2005). What is ecosocialism? *Capitalism Nature Socialism, 16*(2), 15–24.

Mastini, R., Kallis, G., & Hickel, J. (2021). A green new deal without growth? *Ecological Economics, 179*, 1–9.

Mellor, M. (2005). Ecofeminist political economy. Integrating feminist economics and ecological economics. *Feminist Economics, 11*(3), 120–126.

Pollin, R. (2020). An industrial policy framework to advance a Global Green New Deal. In A. Oqubay, C. Cramer, H.-J. Chang, & R. Kozul-Wright (Eds.), *The Oxford handbook of industrial policy* (394–428). Oxford University Press.

Riofrancos, T. (2019, October 24). Plan, Mood, Battlefield: Reflections on the Green New Deal. Contra el Diluvio. https://contraeldiluvio.es/plan-estado-de-animo-campo-de-batalla-reflexiones-sobre-el-green-new-deal/

Shiva, V. (2004). The ecofeminist perspective (Three Texts). *Polis. Revista Latinoamericana, 9.*

Schwartzman, D. (2009). Ecosocialism or ecocatastrophe? *Capitalism Nature Socialism, 20*(1), 6–33.

Schmelzer, M. (2016). *The hegemony of growth: The OECD and the making of the economic growth paradigm*. Cambridge University Press.

Schulz, C., & Bailey, I. (2014). The green economy and post-Growth regimes: Opportunities and challenges for economic geography. *Geografiska Annaler: Series B, Human Geography*, *96*(3), 277–291.

Tironi, M., & Undurraga, B. (2023). Provocations and tensions of the Anthropocene: Towards a geologization of the social. *Estudios Públicos*, *171*, 137–179.

Wallis, V. (2010). Beyond "Green Capitalism". *Monthly Review*, *61*(9), 32–48.

Williams, J. (2024). Greenwashing: Appearance, illusion and the future of 'Green. *Capitalism. Geography Compass*, *18*(1), e12736.

CHAPTER 4

The Just Transition Paradigm and Its Comparative Expression

Joaquín Abarzúa Varela

1 Introduction

The to low-carbon economies presents a major challenge not only in terms of multilevel coordination (local, national, regional, and international) but transition also in managing the demands for inclusion of those directly affected and the distribution of the costs and benefits brought about by socioeconomic transformation. Inadequate management risks "further increasing social inequality, exclusion, civil unrest, and reduced competitiveness of businesses, sectors, and markets" (Lee & Baumgartner, 2022, p. 8).

Given this complexity, it is important for the law to identify the common elements and expressions of such measures to determine whether a particular regulatory expression can be classified as a just transition measure.

To this end, the present work will analyze the various academic conceptualizations of the just transition, seeking to identify their common elements. It will then examine their practical expressions

J. Abarzúa Varela (✉)
University of Chile, Santiago, Chile
e-mail: joaquin.abarzua@derecho.uchile.cl

E. Costa Cordella (ed.), *The Ecological Transition as a Challenge to Law*, Just Transitions,
https://doi.org/10.1007/978-3-032-22055-4_4

in comparative law through the content of nationally determined contributions (NDCs), biennial transparency reports (BTRs), and climate change laws. The aims are (1) to understand whether the various designations of just transition correspond to distinct paradigms or, rather, to typologies within a single conceptual paradigm aspiring to universality; and (2) to contribute to the identification and characterization of those initiatives that can be termed just transition measures.

The multidisciplinary development of the concept of just transition has enriched its content but has also led to interpretative divergences. Thus, the concept has become "broad, multifaceted, and, to some extent, problematically polysemic, leaving room for confusion in interpretation" (Wang & Lo, 2021, p. 2).

It is possible to identify two basic positions regarding the meaning of just transition. On the one hand, a traditional and narrow approach to the term asserts that its labor orientation is fundamental, considering political polarization and "jobs versus environment" tensions (Müllerová, Balounová, Ruppel & Houston, 2023). On the other hand, a broader and more holistic understanding of the concept incorporates aspects of energy, climate, environmental, and social justice (McCauley & Heffron, 2018, as cited in Müllerová et al., 2023).

Nevertheless, the difficulty in determining a universal definition persists. Moreover, its practical translation is equally complex, considering that policies are designed according to the reality of each country, in line with the principle of "no one size fits all" (ILO, 2015).

2 Main Definitions of the Transition

2.1 Typologies

According to the academic literature, there are three main designations when addressing the transition to a low-carbon economy: the labor-oriented just transition; the energy transition; and the ecological or socio-ecological transition or transformation. As can be observed, although the idea of transition is present in all of them, it is also true that they differ considerably in their approach, being concepts that are only weakly connected but that respond to the same historical period of ecological transition. Each of these will be addressed in the following subsections, identifying their main elements.

2.1.1 *Labor-Oriented Just Transition*

The origins of the concept of labor-oriented just transition date back to the 1970s in the United States, in the context of trade union movements. However, these movements were not solely based on labor issues, as there are also records indicating that trade unions had already incorporated an ecological component, decades earlier, denouncing the severe effects of pollution on workers' health (Dewey, 1998). Thus, just transition was used to express their demands in terms of the social and economic conditions necessary for the transition to low-carbon and climate-resilient economies (Pinker, 2020).

Nevertheless, the subsequent connection between trade union movements and environmental justice movements elevated the use of this concept to the international scale, being employed in climate conferences thanks to the influence of the International Labour Organization (ILO) and the International Trade Union Confederation (Thomas, 2021).

The ILO has defined just transition as "greening the economy in a way that is as fair and inclusive as possible for everyone involved, creating opportunities for decent work and leaving no one behind" (ILO, 2022, p. 4). This clearly demonstrates the explicit link between this concept and sustainable development, social inclusion, and social justice, which, together with gender equality and social dialogue, among other concepts, are incorporated within the guiding principles of just transition (ILO, 2023).

2.1.2 *Energy Transition*

The energy transition, in turn, refers to the shift from one dominant energy source to another. Its contemporary use dates to the oil crisis of the 1970s, focusing on dependence on fossil fuels (International Renewable Energy Agency [IRENA] Coalition for Action, 2023), and with a clear objective of economic efficiency (García-García et al., 2020), without an immediate climate or socioenvironmental foundation. Gradually, this concept became linked to the labor-oriented just transition through trade union movements and, in the 1990s, to the climate and environmental crisis, when decarbonization became increasingly relevant for public policy (Geels & Turnheim, 2022).

From these conceptual intersections emerges the just energy transition, understood not only as a change in the production, distribution, and storage of energy but also as a complex array of political and socioeconomic elements to be considered in the transition process (García-García et al., 2020).

This development also incorporated energy justice, understood as "a global energy system that fairly distributes both the benefits and costs of energy services, and features representative and impartial energy decision-making" (Sovacool et al., 2017, p. 1). Although there is no absolute consensus, the principles most commonly associated with the energy transition include (1) flexibility in the face of social complexity; (2) equal access to benefits; (3) participation and consideration of local communities in public policy formulation; (4) ensuring skilled jobs and human resources; and (5) ensuring cooperation among stakeholders and worker participation (García-García et al., 2020).

2.1.3 *Ecological or Socioecological Transition*

Finally, although there are various approaches, the development of the concept of ecological or socioecological transition or transformation as such is often associated with the work conducted by the Institute of Social Ecology in Vienna (Brand & Wissen, 2017). This socioecological transformation can be defined as "the process of changing the way society interacts with the environment in order to reduce socioecological vulnerability and promote sustainable development" (Das Malakar et al., 2023, p. 171)—that is, as a concept that focuses on the relationship between society and nature (Dimitrova et al., 2013).

Its use, however, mainly follows two different orientations. On the one hand, it describes a critical stance toward the idea of just transition, as it obscures the underlying causes of the ecological crisis (Araya et al., 2023), emphasizing power relations, the logic of economic growth, and, ultimately, the modern-colonial extractivist base (Scarpacci, 2023). These orientations are usually linked to principles of climate justice, social justice, ecocentrism, participatory democracy, anti-extractivism, and decolonialism, among others (COPAC, 2022).

The other perspective describes a process similar to the just transition, in that its defining characteristics include, among others, being crosscutting, from the perspective of the actors involved and its multisectoral approach; being just, in the sense of following the idea that "no one is left behind" and seeking to combat inequality; and being participatory, in terms of the horizontal participation of different stakeholders in its design (Jarillo, 2021).

3 Toward a Universal Paradigm of Just Transition

The labor-oriented just transition is the starting point of the concept as we know it today. Likewise, it shares an intrinsic link with the energy transition (Newell & Mulvaney, 2012), as it was the strengthening of environmental governance in the 1970s and its negative effects on employment that led labor movements to develop the concept.

Regarding the socioecological transition, its critical conceptual use posits profound differences in comparison with the just transition. This usage resembles the transformative approach of the just transition model proposed by the Just Transition Research Collaborative, as it entails a comprehensive revision of the existing political and economic system (UNRISD, 2018). However, the same model rejects the existence of distinct transition categories, arguing that these are simply approaches that envision different futures from the perspective of the current political economy (UNRISD, 2018).

Furthermore, it should be noted that the identified definitions share a clear epistemological unity and a common ontology: all conceive reality or are based on similar assumptions (the existence of a climate and ecological crisis) and principles (sustainable development, justice, etc.), albeit with different emphases (labor, energy, productivity, etc.). Ultimately, they all conceive reality in a similar way, as they seek to address the same underlying issue: the dynamic conflict between economic, social, and environmental dimensions, whether it is referred to as the eco-social-economy trilemma (Schoyen et al., 2022), the energy trilemma (Heffron et al., 2015), or the socioecological system trilemma (Das Malakar et al., 2023).

In terms of justice, the academic literature agrees that all typologies of just transition that have been discussed are founded upon and guided by the principles of distributive justice, procedural justice, and restorative justice (Cahill & Allen, 2020; Hoffman et al., 2021). More recently, justice as recognition has also been incorporated, stemming from a broader approach to distributive justice (McCauley & Heffron, 2018).

Therefore, it would not be appropriate to refer to paradigms in the plural sense, as the definitions that are analyzed do not present significant differences in terms of shared systems of principles, techniques, or solutions (Khun, 1970). Rather, it would be more accurate to speak of typologies with close similarities that, in view of this work, can be grouped into a single paradigm of just transition. However, theoretical content alone is not sufficient to determine the real and universal meaning of the concept of just transition. It is also necessary to analyze its comparative expression, which will be addressed in the next section.

4 Expressions at the Comparative Level

At the international level, paragraph 2 of Article 4 of the Paris Agreement (2015) establishes the NDCs as the central instrument through which states report their emissions reduction commitments. The relevance of the NDCs as an object of analysis lies in the fact that the treaty itself states in its preamble that the Parties should take into account "the imperatives of a just transition of the workforce and the creation of decent work and quality jobs, in accordance with nationally defined development priorities" (Paris Agreement, 2015).

The connection between the NDCs and just transition became explicit in the so-called Silesia Declaration (24th United Nations Climate Change Conference [COP24], 2018), which urged states to include issues related to just transition in their respective NDCs.

Equally relevant are the BTRs, as they include, among other things, relevant information on the progress made toward achieving the respective NDCs (UNFCCC, Decision 18/CMA.1., 2019). This is even more significant, considering the declaration "Supporting the Conditions for a Just Transition Internationally" (COP26, 2022), in which the signatory Parties agreed to report on their just transition efforts in the BTRs.

In light of the above, the NDCs and/or BTRs of two countries from each of the five regional groups of United Nations member states, which are also Parties to the United Nations Framework Convention on Climate Change (UNFCCC), will be analyzed. The criteria for their selection are (1) explicit mention of just transition, energy transition, or socioecological transition, whether in their NDC, BTR, or both; and (2) the existence of a climate change law in their domestic legislation, regardless of whether it explicitly mentions the concepts of just transition, energy transition, ecological transition, or socioecological transition.

For the purposes of this analysis, a climate change law will be understood as any comprehensive and multisectoral regulatory instrument that establishes general principles and defines the institutional framework for climate change policy and implementation (World Bank, 2020). The importance of including it as a selection filter, without necessarily analyzing its content, lies in the fact that the presence of legal institutional frameworks promotes effective climate action and, in turn, also facilitates a just transition (Averchenkova et al., 2024).

4.1 *African States: South Africa and Nigeria*

Both South Africa's NDC (Department of Forestry, Fisheries and the Environment, 2021) and its BTR (Department of Forestry, Fisheries and the Environment, 2024) mention various measures related to the energy-oriented just transition, such as the Just Energy Transition Investment Plan (JET IP), focused on electricity, new energy vehicles, and green hydrogen. In parallel, the Presidential Climate Commission (PCC) involves a broad range of societal actors responsible for overseeing the just transition process, ensuring the minimization of costs for workers and communities, and maximizing their benefits. This demonstrates a labor-oriented just transition and reflects the principles of distributive and procedural justice. It is also worth highlighting the Skills Implementation Plan for the Just Energy Transition and the integration of gender, which reinforce these characteristics while incorporating an element of recognition justice.

Finally, Article 1 of its climate change law defines just transition as "a shift towards a low-carbon, climate-resilient, and ecologically sustainable economy and society that contributes to the creation of decent jobs for all, social inclusion, and the eradication of poverty" (Climate Change Act, 2024).

Nigeria's NDC (Federal Republic of Nigeria, 2021) also demonstrates a labor orientation by identifying just transition policies as a prerequisite for climate policies to yield positive labor outcomes. Additionally, its BTR (Federal Republic of Nigeria, 2024) presents an Energy Transition Plan (ETP) and measures focused on the energy and transport sectors. It also mentions a monitoring, reporting, and verification system for an inclusive just and gender transition, expressing components of distributive, recognition, and procedural justice, as it highlights tripartite cooperation among government, workers, and employers' associations to achieve this goal. Lastly, its climate change law will not be addressed here, as it contains no explicit reference to the concept of just transition.

4.2 *Latin American and Caribbean States: Chile and Brazil*

Chile references just transition in its NDC (2020) as a social pillar, highlighting energy-oriented measures such as the "Just Transition

Strategy," which drives the decarbonization of its energy matrix and exemplifies procedural justice by incorporating active public participation in its design and implementation. It also includes labor-oriented initiatives, such as the "Capacity Building and Climate Empowerment Strategy," focused on developing technical skills to promote a just transition for the workforce, thereby expressing distributive justice measures (Ministry of the Environment [MMA], 2020).

Its NDC enhancement (2022) further developed the concept of a Just Socioecological Transition, defining it as "a process that, through social dialogue and collective empowerment, seeks to transform society into one that is resilient and equitable, capable of addressing the social, ecological, and climate crisis" (MMA, 2022, p. 2). This advances a socioecological orientation that includes addressing the ecological crisis as one of its objectives. Subsequently, its BTR (MMA, 2024) confirmed this conceptualization by mentioning the creation of an "Interministerial Committee for a Just Socioecological Transition," an entity responsible for advising and proposing measures to advance the Just Socioecological Transition (MMA, 2022). Finally, it should be noted that Chile's climate change law will not be analyzed, as it does not make any explicit reference to just transition.

In the case of Brazil, its NDC (2024) states that both its national adaptation and mitigation strategies include achieving a just transition as one of their guiding principles. Furthermore, the National Energy Transition Policy establishes the National Energy Transition Plan (Plante) and the National Energy Transition Forum (Fonte) as its main instruments, representing energy-oriented just transition measures (Brazilian Government, 2024).

Its BTR (Ministry of Science, Technology and Innovation, 2024) states that the transition is considered just in the context of sustainable development and efforts to eradicate poverty, affirming that it will take into account socioeconomic inequality, equity, and the reduction of potential impacts on vulnerable groups, including in the workforce transition process. While these are merely statements, they reflect an understanding of a labor-oriented just transition with clear manifestations of distributive, procedural, and recognition justice. Lastly, like Chile, Brazil does not make any reference to just transition in its climate change law, and therefore it will not be analyzed.

4.3 Asia-Pacific States: South Korea and Vietnam

South Korea's NDC (Republic of Korea, 2021) refers to its climate change law, noting, among other things, that it establishes support centers for just transition and the prevention of unemployment, re-employment, and labor retraining. In addition, Article 2, paragraph 13 of this law defines just transition as "a policy orientation to protect workers, farmers, small and medium-sized business owners, etc., in regions or industries that may suffer direct or indirect harm in the process of transitioning to a carbon-neutral society, with the aim of sharing burdens and minimizing harm to vulnerable social groups" (Framework Act on Carbon Neutrality and Green Growth, 2021), reflecting a distinctly labor-oriented understanding and a strong presence of justice principles. Regarding the BTR, there are currently no publicly available records for analysis.

As for Vietnam, its NDC (Socialist Republic of Viet Nam, 2022) presents an energy transition program in the transport sector. Additionally, as part of the development of climate change policies and improvements, it mentions supporting those who lose jobs, creating new employment opportunities, assisting low-income groups when energy prices are high, and protecting women's rights. It should be noted that, like South Korea, Vietnam does not have an available BTR.

Regarding its climate change law, its fourth principle refers to the development of infrastructure that responds to climate change and the energy transition (Prime Minister, 2022). Furthermore, although it refers only to transition without further qualification, it is still useful for the analysis to note its mention of prioritizing training and professional transition in the context of social security measures and gender equity, as well as the participation of women and youth in policy implementation (Prime Minister, 2022). Together with what is already mentioned in its NDC, these elements reflect an approach to a just energy and labor transition, with expressions of distributive, recognition, and procedural justice.

4.4 Western Europe and Other States: Spain and Australia

In the case of Spain, its NDC will not be addressed in this section, as it corresponds to the common NDC of the European Union (EU), which will be reviewed in the context of Eastern Europe. Both its BTR and its Climate Change and Energy Transition Law (Law 7/2021, 2021) display a predominant focus on labor and energy issues, with clear expressions of justice. Through the implementation of the "Just Transition Strategy," the aim is to ensure fair and supportive treatment of workers, minimizing the impacts of the transition on employment (Ministry for the Ecological Transition and the Demographic Challenge [MITECO], 2024). Additionally, the "Just Transition Agreements" are designed to address cases of closure or conversion of facilities by promoting the employability of vulnerable workers and those at risk of exclusion during the transition, as well as regulating the cessation of domestic coal production (Law 7/2021, 2021).

Nonetheless, at least semantically, the BTR also approaches the socio-ecological transition by mentioning "projects in the field of ecological transition" and the use of "nature-based solutions" (MITECO, 2024), although it does not provide concrete examples of what this would entail.

Australia also demonstrates a clear labor and energy orientation. Its NDC (Department of Industry, Science, Energy and Resources, 2022) promotes a just transition fund for the regions, aiming to support the development of new clean energy industries, prioritizing decarbonization and the development of the skills and capacities needed for the transition. The BTR (Department of Climate Change, Energy, the Environment and Water, 2024) confirms this energy transition and highlights the "First Nations Energy Strategy," a notable measure from the perspective of procedural and recognition justice, as it seeks to enable Indigenous peoples to lead and benefit from the energy transition, creating opportunities for effective economic participation and leadership roles (Australia, 2024). Finally, its climate change law will not be analyzed, as it does not refer to just transition.

4.5 Eastern European States: Croatia and Slovenia

Croatia and Slovenia share the EU's NDC and do not have their own BTR, so only the first instrument will be analyzed.

The NDC in question (Spain and the European Commission, 2023) mentions a "Modernization Fund" to support investments aimed at modernizing and increasing the efficiency of energy systems in the 13 lowest-income member states. This is a measure with a clear energy focus and an expression of distributive and procedural justice principles.

Croatia's climate change law (Zakon o klimatskim promjenama i zaštiti ozonskog sloja, 2019) demonstrates a clear labor orientation, as a beneficiary of the EU Modernization Fund. Article 103 states that part of these resources will be allocated to support a just transition, seeking to "provide support for the relocation, training, and retraining of workers, for education, job search initiatives, and for startups, in dialogue with social partners" (Zakon o klimatskim promjenama i zaštiti ozonskog sloja, 2019).

In Slovenia, Articles 143 and 144 of its law *Zakon o varstvu okolja* (2022) regulate and make legally binding the climate change policy instruments, particularly the long-term climate strategy. This strategy refers in its principles and objectives to just transition and also includes a just transition fund.

Furthermore, this climate strategy states that to ensure a just transition, it is necessary to ensure the fair distribution of costs and benefits, as well as to assist the most vulnerable groups of the population and enable them to implement climate change mitigation and adaptation measures (Government of the Republic of Slovenia, 2021). Thus, like Croatia, Slovenia incorporates a legal expression with a strong labor orientation into its conception of just transition.

5 Conclusions

The broad and multidisciplinary nature of just transition has made it difficult to arrive at a universal definition of the term, leading to challenges in its interpretation. Nevertheless, the main definitions of just transition discussed in the literature show more similarities than differences, making it possible to affirm that between just transition itself and its main variants (labor-oriented just transition, just energy transition, and ecological/socioecological transition), there is a genus–species relationship, with all being expressions of the same process and the same idea: the transition toward low-carbon economies, integrating principles of justice (through various approaches), equity, and social inclusion but with emphasis on different elements of their content. Given the existence of a common epistemological and ontological foundation among them, it is more accurate to speak of a single conceptual paradigm that encompasses the different typologies of just transition.

Comparative analysis allows us to affirm the predominance of labor- and energy-oriented just transition typologies, which are often interrelated. Although states may consider them individually in their instruments, these are understood as measures framed within the same transitional process. Furthermore, the principle of justice not only is omnipresent in its classical variants (distributive, restorative, and procedural) but also has a notable emphasis on distributive justice of recognition, through the explicit inclusion of gender issues (in South Africa, Nigeria, and Vietnam) and, exceptionally, matters concerning Indigenous peoples or First Nations (in Australia).

On the other hand, it is worth noting how rarely the socioecological or ecological typology is expressed, as only Chile and Spain refer to it explicitly. However, beyond mere mentions, no differentiated practical expressions are observed at the comparative level. This gap may be due to the identification of this typology with critical perspectives of the status quo as opposed to the "managerial" typologies (labor and energy) preferred by states, which seek to manage the transition within the existing system, thus relegating the socioecological approach to academic and activist spheres for the time being. Nevertheless, these findings warrant more in-depth analysis.

References

Araya, P., Fleischmann, M., Reyes, A., González, K., Oyarzún, T., Sánchez, J., Billi, M., Louder, E., Amigo, C., Urquiza, A., Riquelme, R., & Rojas, V. (2023). *What do we mean when we talk about a just energy transition? Articulating multiple scales, resolutions, and meanings.* NEST-r3 Working Paper No. 4, 16. https://doi.org/10.13140/RG.2.2.28759.65446

Averchenkova, A., Higham, C., Chan, T., & Keuschnigg, I. (2024). *Impacts of climate framework laws: Lessons from Germany, Ireland and New Zealand. Policy Report.* Grantham Research Institute on Climate Change and the Environment, *London School of Economics and Political Science*, *27*. https://www.lse.ac.uk/granthaminstitute/wp-content/uploads/2024/03/Impacts-of-climate-framework-laws.pdf

Brand, U., & Wissen, M. (2017). Social-ecological transformation. *International encyclopedia of geography: People, the earth, environment and technology.* Edition, 3. https://doi.org/10.1002/9781118786352.wbieg0690.

Brazilian Government. (2024). *Brazil's NDC – National determination to contribute and transform.* https://unfccc.int/sites/default/files/2024-11/Brazil_Second%20Nationally%20Determined%20Contribution%20%28NDC%29_November2024.pdf

Cahill, B., & Allen, M. M. (2020). *Just transition concepts and relevance for climate action.* Just Transition Initiative. https://www.csis.org/analysis/just-transition-concepts-and-relevance-climate-action

COPAC [Co-Operative and Policy Alternative Climate Center]. (2022). *People's planning for a deep just transition: Activist guide* (pp. 29–30). https://cjcm.org.za/tools/download/b2ce473a-87ab-467a-95e6-b167693bcca7

Das Malakar, K., Kumar, M., Anand, S., & Kuzur, G. (2023). *Climate change and socio-ecological transformation* (p. 171). Springer Publishing. https://doi.org/10.1007/978-981-99-4390-6

Department of Climate Change, Energy, the Environment and Water. (2024). *Australia's first biennial transparency report.* https://unfccc.int/documents/644997

Department of Forestry, Fisheries and the Environment. (2021). *South Africa's updated first nationally determined contribution (NDC) under the Paris Agreement.* https://unfccc.int/sites/default/files/NDC/2022-06/South%20Africa%20updated%20first%20NDC%20September%202021.pdf

Department of Forestry, Fisheries and the Environment. (2024). *South Africa's first biennial transparency report.* https://unfccc.int/sites/default/files/resource/First%20Biennial%20Transparency%20Report%20of%20South%20Africa%20under%20the%20PA_20%20December%202024.pdf

Department of Industry, Science, Energy and Resources. (2022). *Australia's nationally determined contribution communication 2022.* https://unfccc.int/sites/default/files/NDC/2022-06/Australias%20NDC%20June%202022%20Update%20%283%29.pdf

Dewey, S. (1998). Working for the environment: Organized labor and the origins of environmentalism in the United States, 1948–1970. *American Society for Environmental History*, *3*(1), 45–63. https://doi.org/10.2307/3985426

Dimitrova, A., Hollan, K., Laster, D., Reinstaller, A., Schratzenstaller, M., Walterskirchen, E., & Weiss, T. (2013). Literature review on fundamental concepts and definitions, objectives and policy goals as well as instruments relevant for socio-ecological transition. *WWWForEurope Working Paper*, No. 40, 10. https://hdl.handle.net/10419/125696

Državni zbor Republike Slovenije. (2022, March 29). *Zakon o varstvu okolja (ZVO-2). Uradni list Republike Slovenije, 44/2022.* https://www.uradni-list.si/_pdf/2022/Ur/u2022044.pdf

Federal Republic of Nigeria. (2021). *Nigeria's first nationally determined contribution – 2021 Update.* https://unfccc.int/sites/default/files/NDC/2022-06/NDC_File%20Amended%20_11222.pdf

Federal Republic of Nigeria. (2024). *First biennial transparency report to the UNFCCC.* https://unfccc.int/sites/default/files/resource/Nigeria%20BTR1%20.pdf

García-García, P., Carpintero, O., & Buendía, L. (2020). Just energy transitions to low carbon economies: A review of the concept and its effects on labour and income. *Energy Research & Social Science*, *70*, 2. https://doi.org/10.1016/j.erss.2020.101664

Geels, F. W., & Turnheim, B. (2022). *The great reconfiguration: A socio-technical analysis of low-carbon transitions in UK electricity, heat, and mobility systems* (p. 48). Cambridge University Press. https://doi.org/10.1017/9781009198233

Government of the Republic of Slovenia. (2021). *Slovenia's long-term climate strategy until 2050.* https://unfccc.int/sites/default/files/NDC/2022-06/Slovenia%20LT-LEDS.pdf

Head of State. (2021, May 21). *Law 7/2021, of May 20, on climate change and energy transition. Official State Gazette*, 121, 62011–62060. https://www.boe.es/eli/es/l/2021/05/20/7

Heffron, R. J., McCauley, D., & Sovacool, B. K. (2015). *Resolving society's energy trilemma through the energy justice metric. Energy Policy*, *87*, 168–169. https://doi.org/10.1016/j.enpol.2015.08.033

Hoffman, J., Davies, M., Bauwens, T., Spath, P., Hajer, M. A., Arifi, B., Bazaz, A., & Swilling, M. (2021). Working to align energy transitions and social equity: An integrative framework linking institutional work, imaginaries and energy justice. *Energy Research & Social Science*, *82*, https://doi.org/10.1016/j.erss.2021.102317

Hrvatski sabor. (2019, December 27). Zakon o klimatskim promjenama i zaštiti ozonskog sloja. *Narodne novine*, 127/2019. https://narodne-novine.nn.hr/clanci/sluzbeni/2019_12_127_2554.html

ILO [International Labour Organization]. (2015). *Guidelines for a just transition towards environmentally sustainable economies and societies for all*, *6*. https://www.ilo.org/publications/guidelines-just-transition-towards-environmentally-sustainable-economies

ILO [International Labour Organization]. (2022). *Just transition policy brief: How MSMEs can contribute to and benefit from a just transition*, *4*. https://www.ilo.org/publications/how-msmes-can-contribute-and-benefit-just-transition

ILO [International Labour Organization]. (2023). *Resolution concerning a just transition towards environmentally sustainable economies and societies for all. ILC.111/Resolution V*, *2*. https://www.ilo.org/es/resource/ilc/111/resoluci%C3%B3n-relativa-una-transici%C3%B3n-justa-hacia-econom%C3%ADas-y-sociedades

International Renewable Energy Agency [IRENA] Coalition for Action. (2023). Finding common ground for a just energy transition: Labour and employer perspectives. *International Renewable Energy Agency*. 11. https://www.irena.org/-/media/Files/IRENA/Agency/Publication/2023/Aug/IRENA_Coalition_Just_transition_2023.pdf

Jarillo, A. (2021). *The concept of ecological transition: Paradigm, or right?* European Climate Law Papers, Working paper No. 7/2021, 8. https://doi.org/10.2139/ssrn.3922698

Khun, T. S. (1970). *The structure of scientific revolutions* (2nd ed., p. 175). University of Chicago Press.

Lee, S., & Baumgartner, L. (2022) *How just transition can help deliver the paris agreement.* United Nations Development Programme [UNDP]. https://www.undp.org/publications/how-just-transition-can-help-deliver-paris-agreement

McCauley, D., & Heffron, R. (2018). Just transition: Integrating climate, energy and environmental justice. *Energy Policy, 119*, 11. https://doi.org/10.1016/j.enpol.2018.04.014

Ministry for the Ecological Transition and the Demographic Challenge. (2024). *Spain's first biennial transparency report.* https://unfccc.int/sites/default/files/resource/ESP-BTR-1-%202024.pdf

Ministry of Science, Technology and Innovation. (2024). *First Biennial Transparency Report (BTR1) of the Federative Republic of Brazil.* https://unfccc.int/sites/default/files/resource/BRA_BTR1_2024_ENG.pdf

Ministry of the Environment. (2020). *Nationally Determined Contribution (NDC) of Chile.* https://unfccc.int/sites/default/files/NDC/2022-06/NDC_Chile_2020_espan%CC%83ol.pdf

Ministry of the Environment. (2022). *Strengthening Chile's Nationally Determined Contribution (NDC).* https://cambioclimatico.mma.gob.cl/contribucion-determinada-ndc/descripcion-del-instrumento/#:~:text=presentado%20en%202022-,ac%C3%A1,-La%20actualizaci%C3%B3n%20de

Ministry of the Environment. (2024). *First biennial transparency report and Fifth National Communication to the UNFCCC.* https://unfccc.int/documents/645192

Müllerová, H., Balounová, E., Ruppel, O. C., & Houston, L. J. H. (2023). Building the concept of just transition in law: Reflections on its conceptual framing, structure and content. *Environmental Policy and Law, 53*(4), 276. http://doi.org/10.3233/EPL-230012

National Assembly of the Republic of Korea. (2021). *Act No. 18469, framework act on carbon neutrality and green growth for coping with climate crisis.* https://elaw.klri.re.kr/eng_service/lawView.do?hseq=59620&lang=ENG

Newell, P., & Mulvaney, D. (2012). The political economy of the 'just transition. *The Geographical Journal, 179*(2), 132–140. https://doi.org/10.1111/geoj.12008

Paris Agreement. (2015). UN. https://unfccc.int/sites/default/files/spanish_paris_agreement.pdf

Pinker, A. (2020). *Just transitions: A comparative perspective.* Directorate of Energy and Climate Change, Government of Scotland. https://www.gov.scot/publications/transitions-comparative-perspective/documents/

Prime Minister. (2022). *Decision No. 896/QD-TTg dated July 26, 2022 on approving the National strategy for climate change until 2050*. https://faolex.fao.org/docs/pdf/vie212245.pdf

Republic of Korea. (2021). *Submission under the Paris Agreement: The Republic of Korea's enhanced update of its first nationally determined contribution*. https://unfccc.int/sites/default/files/NDC/2022-06/211223_The%20Republic%20of%20Korea%27s%20Enhanced%20Update%20of%20its%20First%20Nationally%20Determined%20Contribution_211227_editorial%20change.pdf

Republic of South Africa. (2024). *Climate change act, 2024* (*Act No. 18 of 2024*). https://www.gov.za/sites/default/files/gcis_document/202405/50700gen2423.pdf

Scarpacci, M. (2023). From popular energy transition to socio-ecological transition: The city and energy as an opportunity for transformation. *Ecología Política*, (65), 33. https://doi.org/10.53368/EP65TEep02

Schoyen, M. A., Hvinden, B., & Dotterud Leiren, M. (2022). *Towards sustainable welfare states in Europe: Social policy and climate change*. Edward Elgar Publishing, 7. https://doi.org/10.4337/9781839104633.00010

Socialist Republic of Viet Nam. (2022). *Updated Nationally Determined Contribution (NDC)*. https://unfccc.int/sites/default/files/NDC/2022-11/Viet%20Nam_NDC_2022_Eng.pdf

Sovacool, B. K., Burke, M., Baker, L., Kotikalapudi, C. K., & Wlokas, H. (2017). New frontiers and conceptual frameworks for energy justice. *Energy Policy*, *105*, 677. https://doi.org/10.1016/j.enpol.2017.03.005

Spain and the European Commission. (2023). *Update of the NDC of the European Union and its member states*. https://unfccc.int/sites/default/files/NDC/2023-10/ES-2023-10-17%20EU%20submission%20NDC%20update.pdf

Thomas, A. (2021). Framing the just transition: How international trade unions engage with UN climate negotiations. *Global Environmental Change*, *70*, 5. https://doi.org/10.1016/j.gloenvcha.2021.102347

United Nations Framework Convention on Climate Change (UNFCCC). (2018). *Silesia declaration on solidarity and just transition (24th Conference of the Parties)*. https://www.ioe-emp.org/index.php?eID=dumpFile&t=f&f=134978&token=91237abd5b4e38c1e7c2e4364b2b8e7095d8e0fd

United Nations Framework Convention on Climate Change (UNFCCC). (2019). *Report of the Conference of the Parties serving as the meeting of the Parties to the Paris Agreement on the third part of its first session, held in Katowice from 2 to 15 December 2018. Addendum 2. Decision 18/CMA.1*. https://unfccc.int/sites/default/files/resource/CMA2018_3a02S.pdf

UNRISD [United Nations Research Institute for Social Development]. (2018). *Mapping just transition(s) to a low-carbon world*, just transition research collaborative, 12. https://www.uncclearn.org/wp-content/uploads/library/report-jtrc-2018.pdf

Wang, X., & Lo, K. (2021). Just transition: A conceptual review. *Energy Research & Social Science*, *82*, 8. https://doi.org/10.1016/j.erss.2021.102291

World Bank. (2020). *World bank reference guide to climate change framework legislation*. World Bank. 11. https://hdl.handle.net/10986/34972

CHAPTER 5

The Environmental Crisis and Social Contract Theory Within the Framework of the Ecological Transition

Dominga Planella Durruty

Man is born free, and yet everywhere he is in chains. He considers himself master of others but remains more of a slave than they are. How did this change come about? I do not know. What can make it legitimate? I believe I can resolve this question.

Rousseau (1762/2025) (author's own translation)

1 Introduction

Since the beginnings of Western philosophy, there has been an attempt to understand how a group of individuals becomes part of a complex web of relationships, dependencies, and interconnected forms, where the beginning is confused with the end to shape what we call society (Rousseau 1762/2025). This concern is already present in the works of authors such as Plato and Cicero, who referred to the existence of a pact or agreement

D. Planella Durruty (✉)
University of Chile, Santiago, Chile
e-mail: planella.durruty.d@gmail.com

E. Costa Cordella (ed.), *The Ecological Transition as a Challenge to Law*, Just Transitions,
https://doi.org/10.1007/978-3-032-22055-4_5

as the foundation of communal life (Rousseau 1762/2025). Later, what was once a "how" becomes a "why," and the question shifts to the reasons humanity has for organizing itself.

From this arises a political theory that, in its attempt to justify the origin of power, posits the idea of an agreement among individuals, in which subjects relinquish part of their natural freedom to an authority, with the aim of moving from a natural state to a social state and thus guaranteeing various legal values ranging from the protection of life and security to, ultimately, the perpetuation of social order. This notion, known as the "social contract," gives its name to the theory that develops it and constitutes the subject of this chapter.

At present, however, it appears to be one in which the capitalist system inherited from modernity is colliding with planetary boundaries, giving rise to an unprecedented social and environmental crisis. As a consequence, the ecological transition emerges as a scenario of transformation in which humanity must ask itself what role it will play.

Thus, this work seeks to draw a parallel between social contract theory and the ecological transition, with the current environmental crisis serving as both a trigger and a connector between the two. Indeed, this crisis, understood here as a state of nature, would be framed within the process of ecological transition in which humanity can reconsider the foundations of society. This, in turn, presents itself as a culminating action where a new social contract appears as an alternative way to survive the current triple planetary crisis.

To address this issue, an interdisciplinary, comparative, and generalist approach will be adopted in order to examine the intersections between social contract theory, the environmental crisis, and the ecological transition.

2 Theoretical Framework: Social Contract Theory, Ecological Transition, and Environmental Crisis

The theory of the social contract has a long historical trajectory, dating back even to antiquity, where the will of the people was already conceived as the foundation of power. Luis Recaséns, in his work, tells us that Protagoras, from a mythical perspective, considered the existence of a social contract as the basis of laws (Recaséns Siches, 1941). Likewise, the Greek philosophical movement based on the teachings of Epicurus of Samos held that people, through reflection, created the State to obtain certain advantages (Recaséns Siches, 1941). In particular, a contract, or *synzeke*, between individuals was entered into in order for them not to harm one another, since "the formation of the state is one of the most important events that have led humankind from a state of savagery to that of civilization" (Recaséns Siches, 1941, p. 176, author's own translation).

However, it was not until the seventeenth century that classical contractualism emerged with the works of Thomas Hobbes (1588–1679), John Locke (1632–1704), and Jean-Jacques Rousseau (1712–1778) (Teixeira, 2014). These authors, although they differed regarding the scope and content of the state of nature and the rights that were relinquished, posited the idea of a social contract, real or hypothetical, as the foundation of the legitimacy of the State's political power (Teixeira, 2014). This contract entailed the renunciation of certain rights in exchange for the consolidation of the State and the protection it provided (Jaramillo Marín, 2012).

Indeed, classical contractualism describes a presocial state, which, through the agreement of individuals, endows an abstract entity, such as the Leviathan or the State, with sovereignty. Thus, Cortés Rodas observes that, like Rousseau, "both Thomas Hobbes and John Locke use the instruments of the social contract to justify the origin of political power; both start from similar categories: the free, equal, and rational individual in the state of nature, and the pact as the founding element of political power" (Cortés Rodas, 2010, p. 100, author's own translation).

On the other hand, the ecological transition, as previously defined in this work, will be understood for the purposes of delimiting its scope in this chapter as an inevitable movement, without a predetermined direction.

In this regard, following the general line proposed by John W. Bennett, the concept of "transition", broken down into its ecological, technological, sociological, and philosophical spheres (Bennett, 1976), reflects the extensive and complex relationship between humans and their environment. It demonstrates that such a transition is not limited to a reaction to environmental change but encompasses profound transformations in modes of production, social organization, and symbolic representation, reflecting a comprehensive reconfiguration of the interaction between culture (or society) and nature (Bennett, 1976).

Furthermore, Ernst García understands that the ecological transition excludes "any modality of such a process of change that involves degrowth or collapse" (Garcia, 2018, p. 91, author's own translation); that is, it does not have a direction per se and may lead to improvements in socioenvironmental and economic policies, or, conversely, this change could result in greater levels of collapse and breakdown of systems and resources.

Finally, to understand what the environmental crisis is, we must start with the basics: what triggered the alarms regarding planetary boundaries, namely, climate change. In this regard, the United Nations defines climate change as "long-term shifts in temperatures and weather patterns. These shifts may be natural, due to variations in solar activity or large volcanic eruptions. But since the nineteenth century, human activities have been the main driver of climate change, primarily due to the burning of fossil fuels such as coal, oil, and gas" (United Nations n.d., para. 1, author's own translation).

From this, the concept of the "triple planetary crisis" has emerged, comprising the aforementioned climate change, biodiversity loss, and environmental pollution. The significance of this concept lies in the connection between these issues. As Di Paola notes, "these problems cannot be considered in isolation, as they are intrinsically linked and form a threatening triad that has even inaugurated a new geological era, the Anthropocene, marked by the negative impact of humans on our environment" (Di Paola & Asiain, 2023, p. 125, author's own translation).

Moreover, Loorbach perceives the environmental crisis as a symptom of a deeply entrenched form of unsustainable development within the fabric of our societies, referring to these "symptoms of unsustainability" (Loorbach, 2007, p. 12) on a "global scale" such as "the over-consumption of natural resources, social and economic inequalities, loss of biodiversity and climate change problems. At a local level, these symptoms may have a very negative and concrete impact on people's lives: flooding, starvation, poor air and water quality, (armed) conflicts and in general a negative impact on welfare, efficiency and development" (Loorbach, 2007, p. 12).

In this sense, the environmental crisis presents itself as a systemic phenomenon that cannot be analyzed solely from its ecological manifestations; rather, it requires a broader reflection on its impact on the economic, political, and cultural structures of societies.

3 The Environmental Crisis as a Possible State of Nature

In the face of the contemporary environmental crisis, it is worth asking whether this situation can be understood as a return to the state of nature. This hypothesis invites us to revisit the vestiges of ancient conceptions of the origin or transformation of societies, as developed by classical contractarian theory.

To advance this analysis, this chapter proposes breaking down the concept of the state of nature into its constituent elements and examines them in relation to the current environmental crisis. Thus, the environmental crisis could be seen as a contemporary manifestation of the archaic state of nature, revealing structural tensions that challenge the continuity of the modern social contract.

First, one of the most representative features of the state of nature is insecurity, accompanied by chaos and a constant sense of threat. Indeed, the absence of protection against external threats generates a climate of tension, fear, and permanent instability. Hobbes describes this situation as "every man, against every man" (Hobbes, 1909, Chap. XIII), in which each individual has a right to everything, making every person a potential enemy of another, resulting in a situation of "continuall feare, and danger of violent death" (Hobbes, 1909, Chap. XIII), with the consequence that life for man is "solitary, poore, nasty, brutish, and short" (Hobbes, 1909, Chap. XIII). For his part, Locke recognizes that the individual application of natural law, without common mechanisms for resolution, leads to an escalation of violence (Locke, 1764). Each individual acts as a judge in his own case, triggering a cycle of reprisals that perpetuates conflict and threat. Thus, "force without right, upon a man's person, makes a state of war, both where there is, and is not, a common judge" (Locke, 1764, Chap. 3).

Climate change and its threatening and imminent catastrophes, the massive loss of biodiversity, and environmental pollution spreading to water, land, and air, creating a sense of uninhabitability, as well as distrust in technology as a weapon of destruction, as a bomb, or as an epidemic, all generate a widespread sense of impending collapse. It is this environmental tension and fear of the future that define the present moment as a precontractual situation. In this context, García notes that "fear, precariousness, and uncertainty about the future stimulate erratic and short-term responses" (Garcia, 2018, p. 99, author's own translation). According to Toledo, the consequence is that:

> The central circumstance of the contemporary world is a growing concern for the future, as society finds itself immersed in a "gigantic uncontrolled experiment" (McNeill 2000, as cited in Toledo, 2013, p. 42, author's own translation), provoked by itself, where natural and social processes are articulated in an unprecedented way, generating new, unpredictable, and surprising dynamics and synergies that threaten the human species, planetary equilibrium, and all life. This is a crisis of modern or industrial civilization, the solution to which requires new paradigms in all fields of reality, which today has become a complex socio-natural or nature–social system (Toledo, 2013, p. 42, author's own translation).

Both in the contractarian tradition and in the current environmental situation, the absence of effective guarantees in the face of danger constitutes a structural condition of threat. Just as in the work of Hobbes and Locke, insecurity, tension, and fear arise in the absence of a common power to ensure respect for law and life, today the environmental crisis generates a similar perception of collapse and instability.

Secondly, a structural dimension of the state of nature is the lack of a common authority capable of effectively guaranteeing the enjoyment of rights. Although for Locke, rights such as liberty or property exist in the state of nature, their exercise is not possible because of the absence of institutions to ensure their enforcement (Locke, 1764). Locke states that "where there is no longer the administration of justice, for the securing of men's rights ... there certainly is no government left" (Locke, 1764, Chap. 19). Therefore, in the absence of an organized means of enforcing such laws, rights such as property, security, and well-being are easily violated, which implies that without a common power, rights are nothing more than proclamations without coercive force (Locke, 1764). For Hobbes, this situation creates a paradox: a freedom so absolute that it cancels itself out, since no one is obliged to respect the life, property, or word of another (Hobbes, 1909). Rousseau, for his part, introduces a relevant nuance: while he acknowledges that in the state of nature, humans are free and equal, he also warns that such freedom is fragile and vulnerable. The absence of property, laws, and imposed inequalities allows for a kind of original freedom, but this is neither assured nor protected, as he considers that "where the force of laws and the authority of those who defend them cease, there can be neither security nor freedom for anyone" (Rousseau 1755/2012, dedication, author's own translation).

Currently, legal frameworks are insufficient to protect people from environmental uncertainty, and institutional mechanisms are increasingly under strain. In this regard, Erin Ryan notes that "Moreover, even where they find solid legal foundation, both state and federal environmental laws struggle for efficacy in protecting public trust resources" (Ryan, 2025, p. 241) such as water, air, the coast, among others (Ryan, 2025). Thus, environmental rights formally exist in most legal systems but are fragile and easily violated in the context of the relationship between environment and development. As Jaria i Manzano warns, the notion of sustainable development, while it "incorporates the economic growth inherent to the capitalist economy," also "endangers the natural processes on which human well-being is based" (Jaria-Manzano, 2022, p. 460, author's own translation). "Thus, the utopian dimension of the constitutional tradition becomes explicit in the face of the limits posed by the evolution of the Earth System, whose anthropic manipulation cannot ensure the material conditions in which the claim of individual autonomy has developed" (Jaria-Manzano, 2022, p. 468, author's own translation).

Indeed, this dissonance between normative recognition and effective application translates into a loss of institutional security. The planetary crisis, therefore, is not only an environmental crisis but also a crisis of guaranteed protection. As institutions fail to effectively protect fundamental rights in the face of possible scenarios of climate collapse, a structural form of lack of protection is reinstated. As in the state of nature described by the contractarians, the absence of an effective common power prevents the effective enjoyment of rights, making it possible to observe a scenario of generalized vulnerability, characteristic of the state of nature.

Thirdly, the transition from the state of nature to the civil state is not explained solely by the presence of chaos or fear. According to the main contractarian thinkers, this step is due to an accumulation of tensions that destabilize the previous order, making it unsustainable. Each author identifies a different breaking point (insecurity, injustice, inequality), but all agree that when natural conditions no longer ensure common life, it becomes necessary to make a pact. The contract thus appears not only as a means of containing disorder but also as a response to processes that erode a certain original equilibrium.

For Rousseau, the move to the civil state does not arise from chaos but from the emergence of inequality brought about by private property. Property transforms social relations, breaking with the previous situation that was driven by compassion, freedom, and equality (Rousseau 1755/2012). According to his theory, this transition between systems originates in the injustice introduced by the appropriation of the commons, which destabilizes the natural balance. Thus, "it seems that by then things had reached a point where they could no longer continue as they were, since this idea of property, dependent on many previous ideas that could only arise progressively, is not formed all at once in the human mind: processes are needed, to properly acquire industry and knowledge, to transmit and increase them year by year, before reaching the ultimate limit of the natural state" (Rousseau 1755/2012, second part, author's own translation). Therefore, it is not chaos that compels the contract; rather, it is the need to protect that property that generates uncertainty, competition, and inequality (Rousseau 1755/2012).

Hobbes, for his part, locates the origin of the pact in the constant threat arising from unlimited freedom, in which everyone has a right to everything. However, even in his theory, the idea seems to be that at the moment when the Leviathan is destabilized and ceases to protect individuals, they have the right to act as if they were in a state of nature, resulting in the coexistence of a collapsing but still existing system with anarchic spaces. "For the Soveraign, is the publique Soule, giving Life and Motion to the Common-wealth; which expiring, the Members are governed by it no more, than the Carcasse of a man, by his departed (though Immortall) Soule" (Hobbes, 1909, Chap. XXIX). This is because "The Obligation of Subjects to the Soveraign, is understood to last as long, and no longer, than the power lasteth, by which he is able to protect them. For the right men have by Nature to protect themselves, when none else can protect them, can by no Covenant be relinquished" (Hobbes, 1909, Chap. XXI). Thus, the combination of the Leviathan's ineffectiveness and the new rule of the law of the strongest creates that atmosphere of tension, the "law of nature", which is the trigger for the reactivation of the state of nature and, therefore, the renewed search for a new political pact that allows a return to the civil state.

Finally, Locke (1764) understands that in the state of nature, there are rules derived from the law of nature and that when the system cannot guarantee their protection, it begins its process of dissolution. Upon its collapse, the protection of such rights inherently returns to individuals. "Where the laws cannot be executed, it is all one as if there were no laws; and a government without laws is, I suppose, a mystery in politics, unconceivable to human capacity, and inconsistent with human society" (Locke, 1764, (, Chap. 19).

Indeed, people once again become their own judges and executioners, threatening a slide toward a state of war. Thus, upon experiencing the loss of a common superior power, the possibility arises that force may be used against another individual (Locke, 1764).

Today, private property, as a central device in the genesis of inequality, has evolved into new forms of appropriation that are no longer limited to territory but encompass global commons and the very conditions of habitability of the planet (Jaria-Manzano, 2022, author's own translation). As Jaria i Manzano warns, this process occurs insofar as "the separation between the spiritual world and the material world evolves until it offers us the portrait of the abstract individual subject who subjugates matter to satisfy his aspirations for self-determination, which ultimately implies the massive exploitation of natural resources" (Jaria-Manzano, 2022, p. 153, author's own translation).

In this context, what the author calls the "crisis of voracious rationality" (Jaria i Manzano, 2011 p. 153, author's own translation) emerges as a turning point in the foundations of modernity, that is, a contradiction between the principles of rationalization and the emancipation of the modern ideal. In this way, a viability crisis also takes shape because of the limits of capitalist accumulation and technological development (Jaria-Manzano, 2022) and a crisis of legitimacy, since rationalization, once seen as a form of human emancipation, begins to be perceived as a threat to the individual, representing a form of domination associated with consumption and accumulation (Jaria-Manzano, 2022).

The consequences of modernity as the expansion of rationality, technological progress, and capital accumulation, as well as the externalization of production costs onto nature, now face the collapse of the very pillars that sustain them (Wallerstein, 2000). Nature, transformed into an insurmountable physical limit, stands as a barrier to the voracious logic of economic growth, preventing its continuation and putting the supporting structures into crisis. Thus, when the foundations that structured the modern project, faith in progress, the unlimited expansion of knowledge, perpetual economic growth, and the subordination of nature, collapse under the weight of their own contradictions, we face not only an environmental crisis in ecological terms but also a comprehensive civilizational crisis. For what is at stake is not only the stability of ecosystems but also the viability of the way of life that modern societies have built. The environmental crisis, therefore, reveals the structural fragility of our civilization. Indeed, it is not an external accident but the direct result of a logic that has understood nature as an inexhaustible resource and has ignored the biophysical and human limits of its own development. In this sense, to speak of the environmental crisis is, ultimately, to speak of a civilizational crisis.

On the other hand, the breakdown of the modern order is expressed not only in terms of ecological crises or economic inequality but also in what Loorbach calls "persistent problems", that is, "problems in which multiple actors are involved, that occur (differently) on different levels of scale, that are complex and integrated (multi-domain), that require a long-term process to deal with, and that are deeply rooted in our institutional structures. Examples of such problems are unsustainable energy-supply, agricultural, mobility, health-care or education systems" (Loorbach, 2007, p. 318). These problems, such as energy, mobility, agriculture, or health systems, have historically been shaped by criteria of efficiency, specialization, and high resource consumption, which today manifest as multiple symptoms of unsustainability: "energy-crises, pollution, environmental degradation, congestion, health-issues etc." (Loorbach, 2007, p. 318). For this reason, Loorbach argues that rather than addressing the symptoms, transitions must involve a transformation of the basic structures of the social system (Loorbach, 2007).

In conclusion, from the contractualist theory perspective, the transition to civil society arises either from the breakdown of the natural order, caused by inequality, insecurity, and private appropriation, or from the maximum accumulation of tension and fear in the state of nature before its collapse. Indeed, Rousseau, Hobbes, and Locke agree that the social contract emerges as a response to disruptive factors. This logic is reproducible today, insofar as the contemporary environmental crisis reveals the simultaneous erosion of the elements that sustained the modern order, currently constituting a true civilizational crisis. Just as in classical contractualism, the combined erosion of security, equality, and material viability, now reflected in the environmental crisis, marks a turning point that compels a rethinking of the political agreement structuring communal life.

4 The Ecological Transition: The Natural Movement Between the State of Nature and the Social Contract

As we have anticipated, the ecological transition does not have a predetermined direction. It only provides certainty about one thing: change.

This inevitable change means that systems, before being destroyed, attempt to adapt. Smith, Stirling, and Berkhout explain socioeconomic changes through the concept of a "regime" (Smith et al., 2005, p. 1491), understood as a stable and interconnected system of practices, technologies, artifacts, and values that has a specific direction, in which rules, practices, and networks determine the development of the regime (Smith et al., 2005).

An example of a regime would be that of modern agriculture, focused on mechanization and increased use of chemicals, with the aim of boosting productivity (Smith et al., 2005). Similarly, the electricity generation regime has been dominated by rules and practices seeking large-scale centralization, that is, through large, high-capacity power plants (Smith et al., 2005). These regimes are interconnected, forming a network that underpins the practices of an economic system.

However, their stability is not absolute, as they are challenged by what the authors call "pressures" (Smith et al., 2005, p. 1491), that is, both internal and external factors that destabilize the regime and provoke a transformation or change in its direction. These pressures may be political, technological, cultural, or environmental. These are the factors that shape the forces of change, and, in the authors' words: "As the relative balance of selection pressures shifts and adaptive capabilities develop, so the transformation process will change too. The art of governing transitions becomes one of recognizing which context for transformation prevails, and which drivers offer the best leverage for guiding change in a desirable direction" (Smith et al., 2005, p. 1498). Nevertheless, the persistence of regimes is not absolute, as they may become destabilized either because they are unable to adapt on their own or because of a lack of coordination in the adaptability of interconnected systems. Indeed, "governing the ecological transition is like navigating without a compass, at night, in the midst of a great storm. To anyone who points out that the same could be said of any area of government, one can only reply that they should consider the magnitude of this particular challenge" (Garcia, 2018, p. 125, author's own translation).

From this perspective, the ecological transition represents the pressure for change in current systems, as the environmental crisis and the supposed imminence of disaster threaten human life and existence itself. Thus, these pressures affect not only economies but also people's ways of life. Likewise, the ecological transition is not something that any single actor can decide, as governments or economic entities cannot prevent change; they can only coordinate either to adapt to it or, if they lack adaptive capacity, to negotiate a new model of governance, an idea advocated in this chapter. Thus, to quote Wallerstein: "In a short time, we will not find ourselves in a common moment of reconstruction within the framework of the existing system, but rather within a period of transition, of the end of one type of system and the gestation of another" (Wallerstein, 2000, p. 2, author's own translation).

The need for a paradigm shift, toward which the ecological transition compels us, must entail a process of redefining the relationship between society and the environment, that is, the anthropocentric logic reflected in the fragment from Protagoras, "Man is the measure of all things, of those that are insofar as they are, and of those that are not insofar as they are not" (Plato, 1988, p. 147, author's own translation). This transformation implies the transition from one system to another, characterized as a complex, uncertain, and unpredictable process, but also an inevitable one. In the words of Ernst García:

And it must also have a component of transition or social change, with a scope that may range from a minimum, consisting of some substantial modifications to urban forms, transport, and food production, to a maximum represented by a great transformation as the result of a genuine civilizational crisis. In either case, the key word is uncertainty. Extreme uncertainty (Garcia, 2018, p. 95, author's own translation).

Thus, the ecological transition constitutes a broad and diverse phenomenon, which entails a structural transformation of society in the face of multiple challenges. This process involves a reconfiguration of production and consumption patterns, political and social institutions, lifestyles, collective value frameworks, and the fundamental relationship between humanity and the natural environment.

In conclusion, if crisis marks the beginning and the pact its resolution, what lies in between is the transition from one to the other. The inevitable movement, a consequence of the triple planetary crisis, tests systems and exerts pressure on the various domains that sustain them. Thus, depending on the capacity for adaptation or the inability to confront the inevitable collapse of the system, the transition will occur gradually in some areas and abruptly in others. Actors seeking to address it proactively must reflect on what they wish to preserve and, consequently, what they are willing to relinquish.

5 Conclusions

Throughout this work, we have explored how the social contract theory, as formulated by Hobbes, Locke, and Rousseau, can be reinterpreted in light of the environmental crisis, situating humanity in a new state of nature that requires the formulation of a new social pact.

The triple planetary crisis, climate change, biodiversity loss, and pollution, has destabilized the global equilibrium and exposed the limitations of modern economic and social models. The struggle for access to resources, inequalities in environmental costs, and the imminent collapse of the natural environment have created a situation of uncertainty and conflict, akin to the presocial condition described by classical contract theorists.

In this context, the ecological transition emerges as an inevitable process of change that, far from being a mere technological or economic adaptation, represents a profound structural transformation. In this regard, humanity must decide what it is willing to relinquish in order to preserve what it values.

References

Bennett, J. W. (1976). *The ecological transition: Cultural anthropology and human adaptation*. Pergamon Press.

Cortés Rodas, F. (2010). El contrato social liberal: John Locke. *Revista Co-herencia*, *7*(13), 99–132.

Di Paola, M. E., & Asiain, C. (2023). La justicia frente a la triple crisis planetaria. El rol del Derecho Internacional Ambiental. *Revista Del Ministerio Público de la Defensa de la Nación*, *18*, 125–144.

Garcia, E. (2018). La transición ecológica: definición y trayectorias complejas. *Ambienta*, *125*, 86–100.

Hobbes, T. (1909) *Leviathan*. (W. G. Pogson Smith, Ed.). Clarendon Press. (First published 1651).

Jaramillo Marín, J. (2012). Significado e impacto de la noción de contrato social en Rousseau y Kant. Alcances y limitaciones en la teoría democrática. *Civilizar*, *12*(23), 111–124.

Jaria i Manzano, J. (2011). *La cuestión ambiental y la transformación de lo público*. Tirant lo Blanch.

Jaria-Manzano, J. (2022). La insolación de Mirèio. Seis tesis y un corolario sobre los derechos (ambientales) en la era del Antropoceno. *Afduam*, *26*, 449–476.

Locke, J. (1764) *Two treatises of government*. In T. H. A. Millar, et al. (Ed.), (First published 1689.)

Loorbach, D. A. (2007). *Transition management: New mode of governance for sustainable development*. International Books.

Plato. (1988). *Parménides, Teeteto, Sofista, Político*. (M. I. S. Cruz, A. Vallejo, & N. L. Cordero, Trans.). Gredos.

Recaséns Siches, L. (1941). Historia de las doctrinas sobre el contrato social. *Revista de la Escuela Nacional de Jurisprudencia*, *3*(12), 175–198.

Rousseau, -J.-J. (1755/2012). *Discours sur l'origine et les fondements de l'inégalité parmi les hommes*. In *Collection complète des oeuvres*, vol. 1, in -4°. http://www.rousseauonline.ch/Text/discours-sur-l-origine-et-les-fondemens-de-l-inegalite-parmi-les-hommes.php

Rousseau, -J.-J. (1762/2025). *Du contrat social, ou principes du droit politique*. In *Collection complète des oeuvres*, vol. 1, in -4°. https://www.rousseauonline.ch/Text/du-contrat-social-ou-principes-du-droit-politique.php

Ryan, E. (2025). Public trust principles and environmental rights: The hidden duality of climate advocacy and the atmospheric trust. *Harvard Environmental Law Review*, *49*, 226–341.

Smith, A., Stirling, A., & Berkhout, F. (2005). The governance of sustainable socio-technical transitions. *Research Policy*, *34*(10), 1491–1510.

Teixeira, A. V. (2014). Los orígenes filosóficos de la noción de soberanía nacional en el contractualismo político de Thomas Hobbes, John Locke y Jean-Jacques Rousseau. *Revista de Derecho de la Pontificia Universidad Católica de Valparaíso*, *43*, 801–819.

Toledo, V. M. (2013). El metabolismo social: una nueva teoría socioecológica. *Relaciones*, *136*, 41–71.

United Nations. "¿Qué es el cambio climático?" United Nations. https://www.un.org/es/climatechange/what-is-climate-change

Wallerstein, I. (2000). Dilemas del capitalismo contemporáneo, las ciencias sociales y la geopolítica del siglo XXI. *Estudios Sobre Las Culturas Contemporáneas*, *5*(10), 39–60.

CHAPTER 6

Ecological Analysis of Law: Proposal for a Way of Thinking About Law for the Ecological Transition

Sofía Rivera Berkhoff

1 Introduction

The main philosophical approaches to law within the modern paradigm—Hobbes, Kant, and Hegel (Habermas, 2008 [1985])—emphasize the separation between the human and the natural. This exalts the human being as having the highest value because of rationality and will (Latour, 2022). As law reflects and structures social beliefs, and in turn consolidates them (Culliman, 2019), in this philosophical context it acquires the function of guaranteeing individual freedom as the supreme value, without considering nature except in its relation to that freedom (Mangabeira, 1977).

This is compounded by an overvaluation of science, derived from the idea of human superiority based on reason (Hobsbawm, 2001). From this arises legal dogmatics or the science of law, developed during the eighteenth and nineteenth centuries (Ruiz-Tagle, 2013), linked to modern

S. Rivera Berkhoff (✉)
University of Chile, Santiago, Chile;
Heidelberg University, Heidelberg, Germany
e-mail: sofia.rivera@derecho.uchile.cl

E. Costa Cordella (ed.), *The Ecological Transition as a Challenge to Law*, Just Transitions,
https://doi.org/10.1007/978-3-032-22055-4_6

liberal theory that equates dignity with autonomy or freedom, by extracting the human being from nature (Ruiz-Tagle, 2018). It also adopts the mechanistic view, conceiving legal conflicts as isolated compartments, under the premise that individual well-being automatically generates collective well-being (Capra & Mattei, 2015).

This way of thinking proves problematic in the face of the current ecological transition. In this process, the human system and its subsystems—including the economy, politics, and law—must respond to prevent new injustices (Costa, 2025). However, modern law, by perpetuating liberal dogmatic thought, reinforces the very causes of the ecological and climate crisis, so maintaining this structure implies a vicious cycle of injustices.

Therefore, this chapter proposes a new way of thinking about law, through the critical analysis of law. With this methodology, the aim is to deconstruct the modern relationship between human beings and nature, incorporating ecological analysis into social realities and therefore into law.

The first section delves into the philosophical, legal, and political context of modernity, demonstrating how this context does not respond to the ecological transition, with particular emphasis on the limitations of law in this regard. Next, the methodology used to develop a new way of thinking about law—namely, the critical analysis of law—will be justified and explained. On this basis, critical legal analysis will be used to question the prevailing conception of law in the context of the ecological transition. Finally, ecological analysis of law will be proposed as a new way of thinking about law, and its conceptual foundations and guiding principles will be constructed through a literature review.

2 Modernity, Science, and Law

Hegel conceptualizes law as "the existence of free will" (1821, p. 29)—that is, the set of institutions that guarantee human freedom as will (Amengual, 1988). Kant understands it as "the set of conditions under which the will of each can coexist with the will of others according to a universal law of freedom" (1972 [1784], p. x), conceiving it as an instrument for the coordination of freedoms (Ulloa, 2006). In Hobbes (1980 [1651]), civil law arises as a paradox: it secures freedom by means of its limitation (Parra, 2014).

These conceptions reflect the modern worldview. In it, Descartes introduces dualism between subject and world, situating the individual as an independent entity (Giannini, 1981), while Boyle and Hobbes consolidate the mechanistic and empirical vision of science, dismissing the transcendent (Latour, 2022). Thus, the modern dual understanding of human beings and nature as two symmetrically opposed poles is invented, where science is responsible for representing the nonhuman and politics for representing citizens (Latour, 2022). Finally, Hegel (drawing on Kant) understands modernity as a philosophical break from the normativity of the past, with subjectivity and reason established as the supreme tribunal (Habermas, 2008).

Thus, modern thought recognizes the human being as superior to nature because of rationality (Giannini, 1981), since with the loss of the unifying role of religion, reason takes its place (Habermas, 2008). Furthermore, modern science, guided by a *mechanism view*, interprets reality as the sum of its parts, isolating phenomena in order to understand them (Capra & Mattei, 2015).

From this, three essential features arise: (1) the human being is differentiated from nature, which is reduced to an empirical object; (2) reason places the former hierarchically above the latter; and (3) knowledge is obtained through an analytical approach to the parts. This understanding of subjectivity permeates the whole modern culture (Habermas, 2008) and, in law, is reflected in the freedom of will and the aspiration to turn law into a science because of the prestige of reason (Ruiz-Tagle, 2013).

Thus, modern legal science or dogmatics presents itself as objective, when in fact it is laden with myths and ideological positions that especially benefit those in power (Grossi, 2003). This results in an oversimplification: understanding law solely as a norm (Grossi, 2003), which aligns with the mechanistic tendency to subdivide law into individual parts that are judged as independent (Capra & Mattei, 2015). Moreover, modern legal mythology associates law with order and certainty (Grossi, 2003), both elements related to science and modern rationality (Capra & Mattei, 2015).

These modern legal simplifications, combined with the philosophical separation between human beings and nature, underpin the Anthropocene (Staid, 2023). The Anthropocene arises from and is enabled by the narrative of the industrial elites of modernity, who relegate nature to something external and purposeless and the world to a resource, elevating the human as exceptional and progress and freedom as an escape from the limits of nature (Bonneuil, 2015). It is also founded on the idea of progress as overcoming natural limits and on a patriarchal logic of domination (Valera, 2022).

This generates an insurmountable contradiction, since the milestone of the Anthropocene precisely challenges the boundaries, established by modern thought, between nature and culture, and it places those who seek to defend the conception of nature on which our main institutions and sciences have been founded for more than 200 years in a difficult position (Hamilton et al., 2015).

In legal terms, this means that freedom must be reconsidered under the new conditions of instability in our climatic and ecological system, if law is to be able to respond to conflicts in the post-Holocene era (Hamilton et al., 2015). Moreover, the way to address these conflicts cannot be based on the expectation of certainty and objective order, nor on the normative (mechanistic) approach to understanding law, since in the context of climate and ecological crisis, the natural system that underpins law is neither ordered nor rational, nor does it produce certainty (Jaria, 2023). Therefore, it is now necessary to relate law to more appropriate concepts such as complexity, disorder, and conflict, so that it can address the crisis and provide adequate and realistic legal solutions (Jaria, 2023).

However, the modern understanding that has led to the Anthropocene and is incapable of addressing it is not universal; there are other perspectives that conceive the relationship between human beings and nature differently (Staid, 2023). Therefore, in this chapter, we propose to critique the modern legal paradigm through critical legal analysis, and, once this paradigm has been deconstructed, we propose foundations for a new way of thinking about law based on an understanding of the human–nature relationship that enables us to face the ecological transition.

3 Critical Legal Analysis as a Starting Point

Precisely the idea that modern law is highly ideologized (Grossi, 2003) underpins critical legal analysis, which holds that "law seeks to promote certain types of personality, actions, and human values, excluding other forms of thought and action that may be equally valid" (Ruiz-Tagle, 2013, p. 696).

Mangabeira Unger (1977), one of the main authors of critical legal analysis, points out that modern law ultimately generates a dissociation between the modern claim that society is an individualistic civilization in which order and freedom are guaranteed by law, and the modern reality in which interdependence is evident and legal rules play a minor role in social life. Thus, there is privileged treatment of dominant groups, who manage to impose their individualistic vision of society, marginalizing certain social groups, who end up being alienated from the law (Mangabeira, 1977).

This imposition of visions through law not only impacts other human groups but also, in the case of the relationship with nature, affects the entire natural system, through the domination of nature by humans, resulting in the exploitation and destruction of nature with increasingly powerful technologies (Capra & Mattei, 2015). By studying the modern legal discourse and analyzing how it privileges certain human groups and interests, it becomes clear that it excludes and leaves unprotected the commons and communities as forms of relationship; in this sense, what is left out are relationships, since through this mechanism, only individuals are considered (Capra & Mattei, 2015).

This critical analysis is even more imperative in the so-called legal sciences, since from modernity to the contemporary era, the sciences in general have undergone a paradigm shift and have now embraced systems thinking; by contrast, law, in particular, has maintained the biased vision of modernity, retaining this mechanistic perspective of reality (Capra & Mattei, 2015).

Latin America has been characterized as a continent that is home to groups distanced from the dogmatic theory of law and its mechanistic vision; in this sense, Latin American legal criticism emerges as a parallel and related current to the critical studies previously analyzed (Tapia, 2018). This way of thinking about law critically examines modern legal thought, particularly regarding the rationalization of law—a rationalization that equates law with the norm, an element that is especially criticized for being considered by modern thought as an exclusive characteristic of the European Western world (Tapia, 2018). Additionally, this school of thought identifies in modern legal thinking a particular adaptation of legal theory to a method of production and a capitalist economy (Tapia, 2018).

At this point, Latin American legal criticism resembles Latin American political ecology, which holds that the ecological crisis requires an environmental rationality that questions the legitimacy of the paradigms underpinning economic and legal rationality, recognizing other forms of knowledge and wisdom, and the rights of communities ("others") that embody this environmental rationality to reinvent their cultural identities and to shape their ways of life in confrontation and hybridization with modernity (Leff, 2011).

Thus, the methodology of critical legal analysis, when used to examine the relationship between law and nature, aligns and theoretically converges with Latin American political ecology. In both cases, emphasis is placed on otherness and the recognition of diverse paradigms and ways of life under the law. We therefore believe that this is the most appropriate methodology for undertaking a possible ecological analysis of law.

4 The Need to Incorporate Ecological Analysis Into Law

As previously mentioned, the prevailing way of thinking about law fails to address the climate and ecological crisis, and is, in fact, one of the conditions that has led to the emergence of the Anthropocene. A relevant example is the regulation of landownership, as individual property is one of the fundamental elements of modern legal thought (Lobos, 2025). Prior to modernity, the predominant form of managing common goods, such as land, was communal organization, until it was replaced by private property as a consequence of liberal revolutions (Lloredo, 2020). This theory of property, which is based on the fundamental separation between humans and nature, reduces land to an object of appropriation and thus "has the intrinsic limitation of failing to comprehend the complexity of the relationship between humans and things" (Míguez, 2014, §2).

The effects of this shortsightedness regarding the complexity of the soil system generate a series of negative consequences: it creates vulnerability among populations that previously formed part of the communal system, establishes power relations that pressure small farmers for the benefit of large landowners, reduces social cohesion and community security, and, as a result of all of the above, makes the territory and its community more vulnerable to climate change (Murken & Gornott, 2022). In contrast, both studies on the experiences of various communities that promote the collective use of common goods and the Intergovernmental Panel on Climate Change (IPCC) recommend, as an adaptation and mitigation measure to address the effects of climate change, the adoption of a land tenure system based on collective action and traditional knowledge (Murken & Gornott, 2022).

The intrinsic individualism of modern private property leads to the dilemma of the "tyranny of small decisions," devaluing decisions that benefit the collective (Turner, 2022). In this regard, Ostrom (2003) argues that the tragedy that arises in the case of common land tenure, as described by Hardin (1968), would only occur if the owners do not communicate or collaborate, but when owners establish collective networks to care for the commons, the system becomes resource efficient and beneficial for the entire community (Lloredo, 2020). For example, Indigenous communities in Latin America that maintain a traditional relationship with the land they inhabit tend to interact with it according to biological needs and natural metabolism, resulting in its maintenance and sustainable use (Diégues, 2008).

This example shows us how modern law is disconnected from an understanding of the natural environment to which it seeks to apply. Therefore, when faced with the climate crisis, law becomes powerless and fails to find effective and long-term solutions to the situation, something that does occur under other forms of relationship not recognized by modern law and philosophy (Lobos, 2025).

Moreover, the example shows us that law is one of the few areas into which systems theory has not permeated, as evidenced by the fact that the private property system does not comprehend the ecological complexity of the soil system (Míguez, 2025). Systems theory is proposed as a philosophical alternative to the element of modern thought that views reality mechanistically, divided into entities or individual elements (Bertalanffy, 2006), and the human being as hierarchically superior to their environment on the basis of the assumption of being the only rational individual (Gilbert et al., 2021).

This mechanism and hierarchy marked science from the paradigm of classical physics, structuring it analytically with the aim of breaking reality down into ever-smaller isolated units. This way of thinking about science led to the creation of a multiplicity of sciences as subdomains and to the belief that these units were closed systems (Bertalanffy, 2006). The foregoing results in the idea that individuals are autonomous and capable of acting on their own.

This paradigm ignores the symbiotic relationships of living beings, in which the association and interaction of living systems are recognized as the essence of ecology (Gilbert et al., 2021). Thus, for more than a century, natural science has acknowledged that elements of nature can be studied not in isolation but, rather, organically and in terms of their relationships. This gave rise to ecology as a field of study and to systems theory as a mode of thought, both of which posit that ecological communities are interconnected and form networks of relationships (Capra &

Mattei, 2015). Systems are now seen as open, interacting with multiple variables, and their organization is dynamic (Bertalanffy, 2006).

Subsequently, systems theory permeated from the natural sciences into the social sciences, where Luhmann's contribution has been the most significant. This author introduced systems thinking into the social sciences by arguing that society cannot survive as a system of action if the organism does not adapt to an ecological environment (Luhmann, 1996). These findings deconstruct the conception of a separation between culture and nature, as they reveal that culture depends on and is oriented toward the ecological environment—that is, both from and toward nature (Luhmann, 1996).

However, these premises have not yet been adopted by law in general, with regulations and legal relationships remaining rooted in modern frameworks, as exemplified by the case of private property discussed earlier. Yet, in the context of a climate and ecological crisis, which is driving an ecological transition, the law can (1) ignore the issue and continue to deepen the problems associated with modern ways of thinking, or (2) take responsibility and become part of the ecological transition, governing it and establishing parameters to ensure that the transition does not multiply existing injustices (Costa, 2025) through a systemic mode of thought. Clearly, the second option should be chosen, as it ensures our survival, that of future generations, and that of other species.

There is a need, then, for law to take responsibility for and become part of the ecological transition, mitigating the consequences of the Anthropocene and responding to crises, it is necessary to adopt an ecological approach to law that incorporates the advances of systems theory, as has been done in other disciplines.

5 Proposal: Ecological Analysis of Law for Addressing the Ecological Transition

First, clarification is needed. What is proposed here is a way of thinking about law—that is, a legal paradigm that arises from the contributions of various authors who propose a particular vision of law in general (Ruiz-Tagle, 2013). This way of thinking about law is termed "ecological," meaning it relates to ecology as "the study of the reciprocal actions between living systems and their environment" (Turk, 1973). Furthermore, the use of the term "ecological" is intended to distinguish it from "environmental" in the same way that environmental justice, which is based on an anthropocentric paradigm, differs from ecological justice, which incorporates systems theory through an ecocentric paradigm (Montalván, 2020).

Thus, the importance of a paradigm shift is highlighted—from the anthropocentrism of modernity to an ecocentrism justified by the integration of scientific knowledge into moral, political, and legal thought. This is something impossible to consider within modern thought, as previously described. In fact, this way of thinking results precisely in the limitation of human freedom in light of natural boundaries, which challenges one of the pillars of modern thought (Aldunate, 2001). Freedom must be reconsidered under the new conditions of instability in our climatic and ecological system if law is to address conflicts in the post-Holocene era (Hamilton et al., 2015), which justifies the relative loss of freedom in favor of ecocentrism.

An ecological analysis of law, therefore, adopts a perspective that integrates science and morality/politics not as opposing poles but as interacting elements. In this interaction, it draws as fundamental pillars of its paradigm the contributions of systems theory—for example, that nature is the superior system within which humans, along with other living beings, are interdependent components. Thus, human freedom is not absolute but is constrained by the limits of the system and of other individuals within it.

Currently, we can observe some academic proposals aligned with this perspective in law, presenting a critical approach to the modern paradigm and incorporating an ecological and systemic outlook. Below, we will briefly analyze the proposal of Fritjof Capra and Ugo Mattei in their book *The Ecology of Law: Toward a Legal System in Tune with Nature and Community*, published in 2015, and the proposal of José Luis Serrano in his book *Principio de derecho ambiental y ecología jurídica*, published in 2007. These works are chosen on the one hand because they directly refer to a new way of thinking about law (called "ecology of law" or "legal ecology") and, on the other hand, because in this author's view they clearly and directly present the elements that this way of thinking about law should include, which facilitates their presentation here.

Capra and Mattei (2015) state that at the core of the "new ecological legal order" they propose is an anti-mechanistic vision, which holds that social reality is not a sum of individuals but is composed of social networks and communities. They also argue that law should be consistent with and honor the basic principles of ecology, and, in this sense, law is seen not as an independent extract separated from other elements of society but as closely linked and related to politics, economics, religion, and so on.

These authors propose three aspects to be reformed in order to promote ecology of law. First, disconnect law from power and violence. This refers to restoring harmony between human laws and the laws of nature by returning law to community networks, overcoming notions of hierarchy and competition, and incorporating the complexity of relationships between parts and the whole (individuals and their communities).

Next, make communities sovereign. This principle aims for property to be directly held by its owners and for its regenerative use to be protected. To achieve this, it would be necessary to transform capital and technology into commons and for the purpose of the law regulating them to be respected, with the common goal of survival, as applied and interpreted by each community according to its own cultural, economic, and social traditions.

Finally, make property regenerative. In this case, the authors propose that each resource necessary to meet needs should be understood as a common good and therefore governed by the principles of solidarity and regeneration, with the aims not only of preserving current commons but also ensuring the existence of commons in the future.

As can be seen, in a brief and simplified approach, Capra and Mattei's proposal tends to refer primarily to the regulation of property and the power associated with it through law. Reforming these aspects of law may be among the most difficult, and we agree that in practice they are the most decisive elements in generating the impotence of law in the face of ecological transition and in perpetuating the modern worldview.

However, we firmly believe that law is about much more than this, and incorporating a systems theory perspective requires precisely considering all elements and their interrelations. It is here, then, that we must consider the contribution of Serrano (2007). This author indicates that legal ecology is a methodology that sets aside the classical analytical methodology in law to embrace the paradigm of complexity, accepting chance and utilizing dynamism and integrity.

The following principles can be drawn from Serrano's proposal:

1. *The thesis of biconditionality*: This recognizes that law is both a producer and an effect of the ecological crisis. The aim is to develop a legal methodology that incorporates a scientific element and seeks to contribute to other disciplines in addressing the ecological crisis.

2. *Symptomatic reading*: This involves studying the legal system as a symptom of a particular historical form of constructing the human/nature relationship, which involves an exegetical or literal reading, and moving toward a reading that uncovers the privileged interests embedded in the law and determines whether these interests are individual and private or communal and social.
3. *The principle of the insufficiency of law*: The legal system cannot be conceived as a system perfectly balanced with its environment. Thus, for structural ecological transformation, it is necessary to consider law as an open and unbalanced system that nevertheless seeks to remain stable to a certain degree of codification and predictability.
4. *The principle of the necessity of environmental law*: While legal decision-making is only about determining who decides and how, environmental decision-making also concerns whether it is possible to decide on those ecological variables of the natural environment that we know are essential. It assigns to states the primary responsibility for management of system/environment relations and the commitment to ensure they are balanced.

6 Conclusions

This chapter has provided an overview of the conceptual foundations of the ecological analysis of law in order to understand the legal phenomenon, thus laying the groundwork for further theoretical developments on the subject.

An ecological analysis of law, then, appears to follow a clear path in establishing the conceptual foundations for this way of thinking about the legal phenomenon. It begins by recognizing that we are in an ecological transition, moving from one geological era to another (from the Holocene to the Anthropocene), marked especially by a scenario of planetary crises—climate, ecological, water, and biodiversity crises, among others. It identifies as the cause of this transition the force exerted by humans on nature, based on the tenets of modern thought. This paradigm rests on three fundamental pillars: the separation between humans and nature; the hierarchically superior position of the former, granting them control over the latter; and the understanding of life as merely the sum of individual elements. This same cause simultaneously constitutes the obstacle that prevents us from adequately addressing the crises and governing the transition justly. In this context, law appears as one of the few disciplines that has not managed to move beyond the modern paradigm toward a systems theory that would allow it to address the conflicts emerging from the ecological context in which it operates.

Therefore, a critical analysis of law is necessary as a methodology capable of deconstructing the modern legal system and determining its essential elements. However, this methodology itself does not propose a new way of thinking about law; rather, like any methodology, it remains at the level of form and does not provide substance or content.

In this thread of thinking, we find the authors and scholars who have proposed a new way of thinking about law that takes ecological transition as its foundation, critiques the modern paradigm, and also establishes the principles on which law should be based in this context. This way of thinking about law is what we have called the *ecological analysis of law*.

The ecological analysis of law thus proposes, as a principal measure, the incorporation of systems thinking into law, the elimination of human/nature dualism, the recognition of the importance of relationships and communities, and the consideration of a material aspect that determines our management of elements of nature as parts of a whole.

This way of thinking about law is still under development, and perhaps it is always under construction, as long as new scientific knowledge continues to be generated. Just as the Earth system is in constant transformation, social systems must incorporate this idea of continual change into their analyses, and we can only conceive of law as mutable as well, so that its aims align with reality. This, ultimately, is what ecological analysis of law allows us to do.

References

Aldunate, C. (2001). *El factor ecológico: Las mil caras del pensamiento verde*. LOM Ediciones.

Amengual, G. (1988). La filosofía del derecho de Hegel como filosofía de la libertad. Revista Taula (UIB), 10.

Bertalanffy, L. (2006). *Teoría general de los sistemas: Fundamentos, desarrollo, aplicaciones* (J. Almela, Trad.). Fondo de Cultura Económica.

Bonneuil, C. (2015). The geological turn: Narratives of the Anthropocene. In C. Hamilton, C. Bonneuil, & F. Gemenne (Eds.), *The Anthropocene and the global environmental crisis: Rethinking modernity in a new epoch*. Routledge.

Capra, F., & Mattei, U. (2015). *The ecology of law: Toward a legal system in tune with nature and community*. Berrett-Koehler Publishers.

Costa, E. (2025). La inevitable transición ecológica y la deseable transición justa. *Revista del Magíster en Análisis Sistémico Aplicado a la Sociedad*, *52*, 39–58.

Cullinan, C. (2019). *El derecho salvaje: Un manifiesto por la justicia de la Tierra*. Huaponi Ediciones.

Diegues, A. (2008). *El mito moderno de la naturaleza intocada* (6.ª ed.). Annablume. (Obra original publicada en 1996).

Giannini, H. (1981). *Esbozo para una historia de la filosofía*. Editorial Universitaria.

Gilbert, S. F., Sapp, J. Y., & Tauber, A. I. (2021). *Todos somos líquenes: Introducción a una visión simbiótica de la vida* (G. Lankin, Trad.). Hifas Editoriales. (Obra original publicada en 2012).

Grossi, P. (2003). *Mitología jurídica de la modernidad* (M. M. Neira, Trad.). Editorial Trotta.

Habermas, J. (2008). *El discurso filosófico de la modernidad* (M. J. Redondo, Trad.). Katz Editores.

Hamilton, C., Bonneuil, C., & Gemenne, F. (2015). Thinking the Anthropocene. In C. Hamilton, C. Bonneuil, & F. Gemenne (Eds.), *The Anthropocene and the global environmental crisis: Rethinking modernity in a new epoch*. Routledge.

Hardin, G. (1968). The tragedy of the commons. *Science*, *162*, 1243–1248.

Hobbes, T. (1980). *Leviatán, o la materia, forma y poder de una república eclesiástica y civil*. Fondo de Cultura Económica.

Hobsbawm, E. J. (2001). *La era de la revolución, 1789–1848* (F. Ximénez de Sandoval, Trad.). Crítica. (Obra original publicada en 1962).

Jaria, J. (2023). Sistemas supercomplejos y (des)orden constitucional: Las consecuencias del Antropoceno en el derecho. *Catalan Journal of Communication & Cultural Studies*, *15*(2). https://doi.org/10.1386/cjcs_00093_7

Kant, I. (1972). Ideas for a universal history with a cosmopolitan purpose. Editorial Porrúa.

Latour, B. (2022). *Nunca fuimos modernos* (2.ª ed., V. Goldstein, Trad.). Siglo Veintiuno Editores.

Leff, E. (2011). Las relaciones de poder del saber en el campo de la ecología política. En H. Alimonda, C. Toro Pérez y F. Martín (Coords.), *Ecología política latinoamericana: Pensamiento crítico, diferencia latinoamericana y rearticulación epistémica* (Vol. 1). CLACSO.

Lloredo Alix, L. (2020). Bienes comunes. Eunomía. *Revista en Cultura de la Legalidad*, *19*, 214–236. https://doi.org/10.20318/eunomia.2020.5709

Lobos, G. (2025, en prensa). Antropoceno y colonialismo: La exclusión inclusiva del colonialismo en América Latina como condición jurídica de posibilidad para una Nueva Era. *Revista de Justicia Ambiental*, *17*.

Luhmann, N. (1996). *Introducción a la teoría de sistemas* (J. Torres Nafarrete). Universidad Iberoamericana.

Mangabeira Unger, R. (1977). *Law in modern society: Toward a criticism of social theory*. The Free Press.

Míguez Núñez, R. (2014). De las cosas comunes a todos los hombres: Notas para un debate. *Revista Chilena de Derecho*, *41*(1), 7–36. https://doi.org/10.4067/S0718-34372014000100002

Míguez Núñez, R. (2025). *La proprietà ecologica*. Edizioni Scientifiche Italiane.

Montalván Zambrano, D. (2020). Justicia ecológica. Eunomía. *Revista en Cultura de la Legalidad*, *18*, 179–198. https://doi.org/10.20318/eunomia.2020.5272

Murken, L., & Gornott, C. (2022). The importance of different land tenure systems for farmers' response to climate change: A systematic review. *Climate Risk Management*, *35*, https://doi.org/10.1016/j.crm.2022.100419

Ostrom, E. (2003). *El gobierno de los bienes comunes: La evolución de las instituciones de acción colectiva*. UNAM y Fondo de Cultura Económica.

Parra, A. (2014). Hobbes y la paradoja del derecho. *Revista Ciencia Política*, *9* (17).

Ruiz-Tagle Vial, P. (2013). Pensar el Derecho. En P. Grez Hidalgo, J. Wilenmann von Bernath y P. Fuenzalida Cifuentes (Coords.), *Una vida en la Universidad de Chile: Celebrando al profesor Antonio Bascuñán Valdés* (pp. 643–712). LegalPublishing/Thomson Reuters.

Ruiz-Tagle, C. (2018). *Derecho y naturaleza: Marcos legales para la naturaleza y los desastres naturales*. Legge Ediciones.

Serrano, J. L. (2007). *Principios de derecho ambiental y ecología jurídica*. Editorial Trotta.

Staid, A. (2023). *Ser naturaleza: Una perspectiva antropológica para cambiar nuestra relación con el medio ambiente* (M. S. Sairafi, Trad.). Errata Naturae.

Tapia, S. (2018). La crítica jurídica en América Latina: Algunos elementos para la discusión. Nuestrapraxis. *Revista de Investigación Interdisciplinaria y Crítica Jurídica*, *1*(2), 2594–2727.

Turk, A. (1973). *Ecología, contaminación, medio ambiente* (C. Gerhard, Trad.). Nueva Editorial Interamericana.

Turner, K. (2022). The environmental consequences of residential land tenure in single family neighborhoods. *Land Use Policy*, *114*, 105959. https://doi.org/10.1016/j.landusepol.2021.105959

Ulloa, A. L. (2006). El concepto de derecho en Kant. *Letras Jurídicas*, *14*.

Valera, L. (2022). *Human ecology: The ethical challenges of the relationship between humans and the environment* (M. C. Maomed & V. Modafferi, Trans.). Tirant lo Blanch.

CHAPTER 7

The Role of Green Finance in the Private Sector as a Tool for a Just Transition

Ignacia Godoy Tello

1 Introduction

It is undeniable that climate change has placed humanity at a critical juncture—so much so that the Intergovernmental Panel on Climate Change (IPCC) estimates that by the year 2100, if climate policies are not strengthened, average global temperatures could rise by more than 3 °C (IPCC, 2023). Likewise, geologists speak of a new geological age: the Anthropocene, which, as defined by the Oxford English Dictionary, is "relating to or denoting the current geological epoch, regarded as the period during which human activity has been the dominant influence on climate and the environment" (Ellis, 2022).

A crisis, however, can be seen as a turning point: a moment when decisions must be made. Although this is not the first time that humans have influenced environmental degradation, it is now occurring in the context of unrestrained industrialization that has profoundly altered the planet's cycles. The capitalist economic logic, understood as an economic and social system based on private ownership of the means of production

I. Godoy Tello (✉)
University of Chile, Santiago, Chile
e-mail: ignacia.godoy@derecho.uchile.cl

E. Costa Cordella (ed.), *The Ecological Transition as a Challenge to Law*, Just Transitions,
https://doi.org/10.1007/978-3-032-22055-4_7

and the pursuit of profit (Cardoso, 2020), has, in turn, been shaped by neoliberalism and its promotion of the primacy of the free market, in opposition to state intervention, and the opening of the economy to global financial flows (Harvey, 2005).

Neoliberalism is currently the prevailing economic ideology in many countries (Ritzer & Dean, 2011), and the current political–economic system has failed to consider the environmental factor, tending to prioritize exclusively economic efficiency, the maximization of individual profits, and the facilitation of capital flows. In this context, companies have developed and expanded their value chains, shaping society and increasingly distancing producers from consumers, and citizens from experts in ecological matters (Laigle, 2013).

Within the development paradigm promoted by these ideologies, there has been no effective consideration of the environmental impact of operations, nor recognition that economic growth has historically relied on the burning of fossil fuels and the emission of greenhouse gases (GHGs), the main drivers of climate change (Gough, 2023).

In response to the environmental crisis and with the aim of aligning global efforts, in 2015 the United Nations launched the 2030 Agenda for Sustainable Development, with 17 Sustainable Development Goals (SDGs), each with specific targets to be achieved by 2030, whose implementation requires the joint support of governments, civil society, and the private sector. In all scenarios, financing plays a fundamental role, as "finance is the test of our seriousness. Without resources, we simply will not deliver for people and planet. But with adequate, predictable, and sustainable financing, everything is possible" (United Nations, 2019).

Therefore, advancing toward the SDGs requires ensuring the necessary financial and political support, in line with sustainable development. It is a fact that the ecological transition is underway and that financing models must adapt accordingly. The notions of "business as usual" or "finance as usual" can no longer be sustained. At this point, green finance in the private sector has become a key player.

This chapter will address the thesis that green finance can be an ally in the ecological transition, particularly for companies, and may become the main driver of change for the private sector to contribute to solutions to the environmental crisis, in line with environmental, social, and governance (ESG) criteria and in pursuit of achieving the goals set out by the SDGs.

2 Finance as a Tool for a Just Ecological Transition

The financial sector plays a crucial role in economic development (Afzal et al., 2022). Among its characteristics, we find that (1) it produces ex ante information on potential investments and capital allocation; (2) it facilitates negotiation, diversification, risk management, and the exchange of goods and services; and (3) it mobilizes and pools savings (Levine, 2005).

There are international agreements that set targets to address the environmental degradation caused by an excessive development model. One of the most significant is the Paris Agreement and the limits it establishes regarding the increase in average global temperature (UNFCCC, 2015). Achieving this outcome requires a series of actions financed by multilateral public funds, such as the Climate Investment Funds (CIF), an alliance of six multilateral development banks, aimed at channeling financing for activities that support climate action (Climate Investment Funds, 2025).

Without undervaluing the importance of such international public agreements, this section focuses on private financing, considering the specific perspective of companies or corporations. This is because companies commonly use credit mechanisms to finance their operations, and, given the critical situation we face, it is not enough to focus solely on the public sector, which (often because of lack of agreements or other delaying factors) may lag behind the constant and rapid progress of global warming. A clear example of this was the unsatisfactory outcome of the 29th Conference of the Parties to the United Nations Framework Convention on Climate Change (COP29), whose primary goal was to increase climate finance funds to US$1 trillion by 2025 but managed to secure only US $300 million annually through 2035 (Global Compact, 2024).

Certainly, the signing of international agreements enables states to improve their ability to set guidelines and access funds for large-scale projects. However, it is also important to consider the impact of investments that companies can make on their own to help achieve global environmental objectives. In this regard, many industries are still in the early stages of developing a sustainability strategy and require long-term investments (Kim, 2018), highlighting the importance of a targeted approach.

Regarding the type of financing aimed at this purpose, green finance has been described as "ecological growth and the transition to a green economy, while reducing negative impacts on the environment" (Gharleghi et al., 2024).

However, rather than a definition, this is a characterization of what green finance seeks to achieve. Some define sustainable finance as "financing to support sectors or activities that contribute to achieving or improving at least one of the relevant dimensions of sustainability" (Migliorelli, 2021). Others describe it as "the process of taking environmental, social, and governance (ESG) considerations into account when making investment decisions in the financial sector" (Brühl, 2022).

For the purposes of this chapter, we will not make a conceptual distinction between green and sustainable finance, understanding that, in general terms, what makes financing green or sustainable is the purpose or justification of the funds. It is essential that the projects and portfolios that are financed have a positive environmental impact, which may include clean energy projects, water use reduction, or initiatives aimed at efficient resource use, among others. We will focus particularly on financing from private banking entities, non-bank institutions, and investment funds.

One of the essential components for an ecological transition is the existence of capital flows toward low-carbon and sustainable projects from traditional sectors (Gehring & Parot, 2016), where banks and investors play a leading role in this capital flow, which, in turn, aims to encourage companies to join the transition. Private financing will be responsible for mobilizing capital within companies, with investors able to influence the companies in which they invest, thereby controlling and directing corporate portfolios. This plays a key role in advancing a just ecological transition, as financing can enable companies to prioritize the development of sustainable projects (Schoenmaker & Schramade, 2019). Thus, the role of green finance would be to guide investments toward more sustainable projects that are also attractive for private actors to participate more effectively in the transition.

From the perspective of companies, there are different types of green financing that vary in how they operate and the requirements for their issuance. For the purposes of this chapter, I will briefly analyze three types of green financing that may be attractive to companies that already have a project or strategy aimed at mitigating or adapting to climate change.

2.1 Green Bonds

Conventional bonds are securities that represent a debt between the bond issuer and the purchaser. The advantage of using a bond is usually to finance a long-term project, allowing them to be bought and sold on the stock market.

The main difference between conventional bonds and green bonds is that the latter include a "clause" specifying their use and purpose: to finance green investments. Therefore, they can be defined as conventional fixed-income assets issued to finance projects with environmental or climate benefits (Soe et al., 2019). This identification entails a commitment on the part of the purchaser to use the proceeds from the bond issuance for projects that are aimed at climate change mitigation or adaptation, or that are beneficial to the environment. For this reason, they may be attractive to private entities seeking to finance projects that fall within this category.

These fixed-income instruments offer a range of benefits for the issuer, including the possibility of attracting more investors and conveying a message of environmental concern to stakeholders regarding sustainability (Bachelet et al., 2019). For the purchaser (i.e., the company), the benefit may lie in the higher profitability of green bonds and the opportunity to demonstrate a contribution to corporate social responsibility (CSR).

This type of green financing is governed by voluntary guidelines known as the Green Bond Principles (ICMA, 2015), which aim to promote transparency, integrity, and sustainability in the green bond market. Since 2013, the private sector has become increasingly active in this trend (Weber & Vasundhara, 2019), experiencing exponential growth. In fact, in 2019, green bond issuance surpassed US$100 billion for the first time (Deschryver & de Mariz, 2020).

2.2 Bilateral Green Loans

Bilateral green loans follow a logic quite like that of green bonds, being granted for projects with environmental objectives. Among the key differences, bilateral green loans, as their name suggests, consist of a loan agreement regulated between only two parties: the lender and the borrower. In this way, they can establish terms specific to the agreement, which do not necessarily match market interest rates or maturities, and they are not transferable in the market.

Bilateral green loans are governed by a set of principles known as the Green Loan Principles (LMA, 2023), which provide guidelines for the borrower when applying for the loan and follow the same logic as the Green Bond

Principles (BBVA, 2018), applicable to green bonds; that is, they establish four characteristics that define a green bond or loan: (1) use of proceeds; (2) project evaluation and a selection process; (3) management of proceeds; and (4) reporting.

2.3 *Sustainability-Linked Financing*

This type of financing qualifies as green because in order to be requested by a company, the latter must have a key performance indicator (KPI) related to a sustainability target of the business. Notably, the funds do not necessarily have to be allocated to that target and may be used for any other corporate purpose, such as payroll expenses, provided that the specified KPI meets the target.

As this is a form of financing linked to sustainability, the KPI justifying the request may be social (increased hiring based on gender, LGBTQ +, or disabilities) or environmental (reduction of GHG emissions, efficient energy use, etc.).

This type of financing, like those mentioned above, is governed by voluntary guidelines or principles known as Sustainability-Linked Loan Principles (LMA, Loan Market Association (LMA), 2025). These guidelines differ from those already mentioned for green bonds and loans in their features, as they specify five components, namely (1) KPI selection, (2) sustainability targets, (3) loan characteristics, (4) reporting, and (5) external verification of KPI compliance.

3 Benefits of Using Green Financing Instruments From a Business Perspective

On the basis of the above review, we can conclude that green financing can be highly attractive for a company with sustainability-related goals or in need of funding to develop a green project.

From a business perspective, having access to this type of financing entails multiple benefits for companies, as it not only involves lower interest rates or fees for their projects or targets but also serves as a signal of CSR (Deschryver & de Mariz, 2020), which, in turn, leads to improved access to financing. CSR can be viewed from different perspectives and, in fact, has not always been well regarded within the economic paradigm, as the neoclassical view considers it "unnecessary and inconsistent" (Liang, Liang & Renneboog, 2020) with the ideals of neoliberalism—that is, with the pursuit of maximum capital gains. Others, on the other hand, have argued for the importance of reaching agreements on the implementation of CSR, which could contribute to sustainable development (Monzón & Morales, 2024). As previously discussed, the purely neoclassical view has, one might say, necessarily been left behind. CSR was the starting point, as ESG criteria are now understood to be included.

Currently, ESG criteria are an important economic factor that frames the social and environmental outlook of companies, and their observance can contribute to the ecological transition by attracting investments that finance projects aimed at reducing environmental impact, thereby supporting climate change mitigation and adaptation, and potentially turning such projects into a fiscal incentive. On the other hand, the primary objective of CSR reports or ESG criteria is to raise transparency standards regarding to social and enviromental impact of companies and their governance structure, with the potential of lead to changes in internal controls that ensure regulatory compliance and the reliability of reporting (Cheng et al., 2014).

Similarly, Flammer conducted a study in 2019 on the effectiveness of green bonds in environmental impact, which is, in turn, related to the ecological transition. The conclusions indicated that corporate green bonds provide several benefits, both to companies and to the environment, representing a powerful tool for financing the fight against climate change (Flammer, 2021), which aligns with the idea of green finance as a driver of the ecological transition.

Currently, private sector investments that support the achievement of any SDG target (targets must be specified when requesting green financing from a bank) remain low, accounting for only a fraction of the assets invested by companies in pursuit of the 2030 Agenda. However, this type of investment was, until recently, on the rise (Global Compact, 2022).

4 Risks and Challenges in the Use of Green Finance as a Mechanism for a Just Ecological Transition

If we understand that the ecological transition is the process of shifting to a more sustainable paradigm, and that it is inevitable, it is worth questioning how much influence green finance can truly have. This is where certain complexities arise.

Firstly, as previously noted, there is no universal definition of what is considered "green." For example, the issuer determines what will be considered green in a bond or loan and may (but is not required to) follow voluntary guidelines. This raises the question: can we truly define green finance as genuinely green? Furthermore, there is no institutional framework to ensure effective transparency regarding what is considered green in financing and the monitoring of statements (Forest, 2023) made by companies seeking such financing, as this monitoring usually depends on external consultants or, in some cases, on the company itself.

The absence of a single concept and lack of transparency entails an additional risk: greenwashing (Kocher, 2019). The lack of a common understanding, left to the discretion of the relevant parties, allows projects to be labelled as "green" without having a real environmental impact. This becomes a particular risk if the country where the project is carried out lacks greenwashing regulations, relying solely on good faith that the company is being truly transparent, with no associated penalties if investments are not used for their stated purpose (Deschryver & de Mariz, 2020).

On the other hand, from a different perspective, it should be noted that not all companies will be able to access this type of financing, as small and medium-sized enterprises usually do not have the same conditions or advantages as large companies when dealing with international investors. Companies that prioritize ESG criteria will likely have a competitive advantage in accessing green finance from banks (Fu and Pirabi, 2023). In contrast, it is rare for a small or medium-sized company to have a sustainability strategy or ESG perspective, as their business focus is likely to be more narrowly centered on the production of goods or the provision of services.

Likewise, for green finance to enable a company to develop green projects, it usually entails a high cost or significant short-term investment with long-term benefits, which may pose a risk to the company's development and, therefore, may not serve as a motivating factor for change. In fact, according to a survey by the United Nations Global Compact, 41% of large companies in Spain have green financial instruments, while among small and medium-sized enterprises, this figure drops to 15% (Global Compact, 2022).

Moreover, not all companies operate in countries with equal conditions. In this regard, developing countries may face limitations or lack the capacity to transition, where, even if a company wishes to contribute to the process, the methodology used to promote a project may not take into account the real impact or risk (Amin et al., 2014).

Finally, from a local perspective, and considering that the ecological transition should ultimately benefit people, there is no guarantee that investment in "green" projects will have any positive impact on the communities where they are located or on the people working in those companies.

4.1 *Main Challenges of Green Finance as a Transition Tool*

4.1.1 *Role of Companies*

As discussed, companies have a role that can be crucial in the ecological transition. Therefore, the business sector as a whole must be aware that it can play an active, key, and indispensable role. As mentioned, the transition is underway, so senior management should understand the positive impact of sustainability both in their business operations and in joint efforts to transition (Global Compact, 2022).

To advance this business role, it is necessary to promote a paradigm shift in business strategy, making it more sustainable and understanding, at the same time, that sustainable finance will serve as a tool for resilience in times of crisis.

It is also important that smaller companies or projects have equal access to green financing. To achieve this, banks must open up to such projects, offering preferential rates or benefits according to the quality of the project.

4.1.2 *Lack of Clear Regulation and Transparency*

The problem of the lack of binding regulations governing green finance means that we do not have a single, unified concept of green finance, nor is there a consistent definition for green finance categories. The absence of clear regulation not only creates potential conceptual confusion but also prevents us from determining the scope of what is considered "green." In the words of Guamán (2022), "private sustainability standards are filling an institutional void."

One tool that helps address this issue is green taxonomies. These allow for the classification and identification of activities that contribute to sustainable development (Kuzmin & Alimov, 2024), using science-based metrics and criteria (Brühl, 2022). This makes it possible to determine which objectives possess the necessary characteristics to be considered genuinely "green" or "sustainable" when seeking green financing.

The development of green taxonomies is the responsibility of states. The European Union is a pioneer in creating a green taxonomy, and several countries have developed their own taxonomies following European criteria (Gondijan & Merle, 2022).

With a clear regulatory framework, it would even be possible to foster innovation in sustainable finance, allowing banks to offer new types of green financing with well-defined boundaries. This, in turn, serves as an incentive to align financial performance with ESG criteria (Kocher, 2019).

The lack of an adequate transparency standard inevitably leads to risks of greenwashing. To address this shortcoming, some propose that companies develop an internal culture of transparency (Kocher, 2019); however, We consider this insufficient. This chapter proposes that states, through legislative power, promote bills to regulate greenwashing and include such measures in these initiatives. In other words, regulations should not only view greenwashing as a risk of misleading advertising to consumers but also encompass green finance.

4.1.3 Business Perspective Versus Community Perspective

Green projects that receive financing generally align solely with business perspectives and objectives. While, from a general standpoint, this represents progress toward environmental protection (through measures such as reduced water use, decreased pollutant waste, or lower GHG emissions), questions remain regarding local issues. This is because the needs of the company do not necessarily correspond directly to the views of the local community. Consider, for example, a company that, in order to operate, uses a large amount of water in an area where, although there is no water scarcity, the amount of water available to the community decreases each year. This same company, through a green project, seeks to reduce the amount of waste it generates, without addressing water use. In such cases, waste reduction will benefit the community, but what about water? There is no requirement for the technical aspects or eligibility of the bond or loan to be related to the specific location where it is implemented.

Although the example above is merely illustrative, similar situations can actually occur. This is where regulation becomes necessary, whether internally within the private sector (by modifying the principles governing green finance) or through binding legislation, standards, and guidelines enacted by legislative authorities. For the ecological transition to be truly just, it is not enough to make progress by focusing on only one factor while neglecting others, such as communities. Therefore, green projects should always consider all areas of impact, and there should be an obligation to take the area of influence into account.

5 Conclusion

A just ecological transition requires reconciling economic development with environmental protection. At this point, the main conflict arises between profit maximization and ecosystem preservation. Although public sector economic efforts have been made through certain financing agreements, this chapter invites a review of the options available to private actors—key players in the prevailing economic model—to facilitate ecological transition. In this way, three financing instruments have been analyzed, outlining their advantages and disadvantages from the perspective of the transition, highlighting, among the main challenges, the risk of greenwashing, the lack of effective regulation (with only guidelines in place), the absence of a real transparency obligation, and the business versus community perspective.

Green finance serves as a strategic driver to redirect the economic model toward sustainability, provided it is accompanied by clear regulatory frameworks, robust international standards, and mechanisms that ensure tangible benefits for communities, thereby guaranteeing that the transition is both environmentally effective and socially equitable.

References

Afzal, A., Rasoulinezhad, E., & Malik, Z. (2022). Green finance and sustainable development in Europe. *Economic Research-Ekonomska Istraživanja*, *35*(1), 5150–5163. https://doi.org/10.1080/1331677X.2021.2024081

Amin, A. L., Dimsdale, T., & Jaramillo, M. (2014). Challenges for green finance. In *Designing smart green finance incentive schemes: The role of the public sector and development banks* (pp. 6–7). E3G. https://www.jstor.org/stable/resrep17714.6

Bachelet, M., Leonardo, B., & Manfredonia, S. (2019). The green bonds premium puzzle: The role of issuer characteristics and third-party verification. *Sustainability*, *11*(4), 1098. https://doi.org/10.3390/su11041098

BBVA. (2018). *Green loans and green loan principles*. Available at https://www.bbva.com/en/green-loans-and-green-loan-principles/

Brühl, V. (2022). Green financial products in the EU: A critical review of the status quo. *Intereconomics*, *57*, 252–259. https://doi.org/10.1007/s10272-022-1057-2

Cardoso, P. (2020, December). The unbearable lightness of "green" *Capitalism. E-cadernos CES*, 34.

Cheng, B., Ioannou, I., & Serafeim, G. (2014). Corporate social responsibility and access to finance. *Strategic Management Journal*, *35*, 1–23. https://doi.org/10.1002/smj.2131

Climate Investment Funds. (2025). *CIF Funding*. Available at https://www.cif.org/cif-funding

Deschryver, P., & de Mariz, F. (2020). What future for the green bond market? How can policymakers, companies, and investors unlock the potential of the green bond market? *Journal of Risk and Financial Management*, *13*(3), 61. https://doi.org/10.3390/jrfm13030061

Ellis, E. C. (2022). *The Anthropocene: A brief introduction*. Alianza Editorial.

Flammer, C. (2021). Corporate green bonds. *Journal of Financial Economics*, *142* (2), 499–516.

Forest, N. (2023). Green bonds: A comprehensive analysis. *Latin American Journal of Economic Development*, *21*(special), 99–130. P 108. https://doi.org/10.35319/lajed.2023esp308.

Fu, C., Lu, L., & Pirabi, M. (2023). Advancing green finance: A review of sustainable development. *DESD*, *1*, 20. https://doi.org/10.1007/s44265-023-00020-3

Gharleghi, B., Shafighi, N., & Nawaser, K. (2024). Green finance and its role in sustainability in the EU. *Journal of Economy and Technology*, *2*, 208–215.

Gehring, M., & Parot, G. (2016, October). The legal transition to a green economy. *Environmental Law Review*, *6*, 8–43. https://revistaderechoambiental.uchile.cl/index.php/RDA/article/view/43314

Global Compact. (2022). *Sustainable finance and the 2030 agenda: Investing in the transformation of the world*.

Global Compact. (2024). *COP29: Conclusions from a summit marked by climate finance*. Available at https://www.pactomundial.org/noticia/cop29-financiacion-climatica/

Gondijan, G., & Merle, C. (2022). *Sustainable taxonomy development worldwide: A standard-setting race between competing jurisdictions*. https://gsh.cib.natixis.com/ourcenter-of-expertise/articles/sustainable-taxonomy-development-worldwide-a-standard-setting-race-between-competing-jurisdictions

Gough, I. (2023). *Global warming, greed and human needs: Climate change, capitalism and sustainable well-being*. Miño y Dávila.

Guamán, A. (2022). Sustainability, climate change and due diligence mechanisms: Challenges and opportunities. In *The fight against climate change and the recognition of the rights of nature: Synergies in Mediterranean cooperation.* In *Proceedings of the conference held at the University of Valencia on June 29–30, 2022.* Piraeus.

Harvey, D. (2005). *A brief history of neoliberalism.* Oxford University Press eBooks. https://doi.org/10.1093/oso/9780199283262.001.0001

ICMA. (2015). *Green bond principles: Voluntary process guidelines for issuing green bonds.*

IPCC. (2023). Summary for policymakers. In H. Lee & J. Romero (Eds.), *Climate change 2023: Synthesis report: Contribution of working groups I, II and III to the sixth assessment report of the intergovernmental panel on climate change IPCC.* p. 11. https://doi.org/10.59327/IPCC/AR6-9789291691647.001

Kim, J. (2018). Leverage the financing role of banks for low-carbon energy transition. In V. Anbumozhi, K. Kalirajan, & F. Kimura (Eds.), *Financing for low-carbon energy transition* (p. 189–210). Springer eBooks. https://doi.org/10.1007/978-981-10-8582-6_8

Kocher, I. (2019). Green finance, at the heart of the ecological transition. *Banque de France Financial Stability Review, (23), Greening the financial system: The new frontier.*

Kuzmin, E., Mirzaev, B., & Alimov, U. (2024). Editorial: Green taxonomy for sustainable development. *E3S Web of Conferences*, 574, 00001. https://doi.org/10.1051/e3sconf/202457400001

Laigle, L. (2013). For a societal-oriented ecological transition. *Mouvements*, *75*(3), 135–142. https://doi.org/10.3917/mouv.075.0135

Liang, H., & Renneboog, L. (2020). *Corporate social responsibility and sustainable finance: A review of the literature.* European Corporate Governance Institute, Finance Working Paper No. 701/2020. Oxford Research Encyclopedia of Economics and Finance. Available at https://ssrn.com/abstract=3698631 or https://doi.org/10.2139/ssrn.3698631

Loan Market Association (LMA). (2023). *Asia pacific loan market association (APLMA) and loan syndications and trading association (LSTA).* London Green Loan Principles LMA.

Loan Market Association (LMA). (2025). *Asia Pacific Loan Market Association (APLMA) and Loan Syndications and Trading Association (LSTA).* Sustainability-Linked Loan Principles LMA.

Levine, R. (2005). Finance and growth: Theory, mechanisms and evidence. In P. Aghion & S. N. Durlauf, *Handbook of economic growth* (pp. 865–923). Elsevier.

Migliorelli, M. (2021). What do we mean by sustainable finance? *Sustainability*, 1–17.

Monzón, R., & Morales, A. (2024). *Investments in ESG companies: Fundamentals and evaluation.* Ecoe Ediciones.

Ritzer, G., & Dean, P. (2011). *Globalization: The essentials.* Wiley-Blackwell.

Soe, A., Xie, H., Luke, B., & Horan, K. (2019). *A look inside green bonds: Combining sustainability with fixed income*. Available at https://www.spglobal.com/spdji/es/documents/research/research-a-look-inside-green-bonds-es.pdf

Schoenmaker, D., & Schramade, W. (2019). *Principles of sustainable finance*. Oxford University Press.

UNFCCC. (2015). *Paris agreement*. Available at https://unfccc.int/sites/default/files/spanish_paris_agreement.pdf

United Nations. (2019). *Secretary-General's remarks at high-level dialogue on financing for development*. https://www.un.org/sg/en/content/sg/statement/2019-09-26secretary-generals-remarks-high-level-dialogue-financing-for-development-delivered

Weber, O., & Vasundhara, S. (2019). *The growth of green bonds. green bonds: Current development and their future*. Centre for International Governance Innovation. Available at https://www.jstor.org/stable/resrep24966.12

CHAPTER 8

International Energy Law in the Age of Transition: Realigning Regulation Towards Sustainability?

Felipe Pino Zúñiga

1 Introduction

Over the past decade, Energy Law has established itself as a quasi-independent discipline within Public Law. It is often compared to Environmental Law, as a specific legal field that, while intrinsically linked to Administrative Law, possesses certain inherent characteristics that allow for its distinction and identification even across different legal systems. At the same time, it is understood that, as with environmental matters, the management of energy issues—whose effective governance depends on the combination of various disciplines and a technical–political approach—could benefit from a special administrative regime.

Specialized scholarship has ventured various definitions of Energy Law, with one of the most widely recognized being that of Heffron and Talus (2016), who define it as "the regulation of the obligations and rights related to energy, of the various stakeholders, throughout the energy

F. Pino Zúñiga (✉)
University of Chile, Santiago, Chile
e-mail: felipe.pino@derecho.uchile.cl

E. Costa Cordella (ed.), *The Ecological Transition as a Challenge to Law*, Just Transitions,
https://doi.org/10.1007/978-3-032-22055-4_8

cycle." This descriptive definition is often complemented by others that add a specific telos, referring to "the rules to ensure that the final supply of energy to users is safe, economically efficient, and environmentally sustainable" (Del Guayo Castiella, 2020).

However, the increasing globalization of energy matters, particularly with regard to international markets for energy commodities (mainly fossil fuels and, more recently, "critical minerals" and "green hydrogen"), as well as international cooperation agreements between States to promote the energy transition to renewables, has raised the question of the existence of an International Energy Law (IEL) as an independent subdiscipline of Public International Law (Roeben & Mete, 2020).

This chapter does not aim to confirm or refute the necessity and relevance of such a subregime; rather, assuming the analytical utility of this category and considering the considerable—both positive and negative—impacts that the global energy transition is projecting on countries of the Global South, it will explore the potential of International Law in the Latin American energy transition.

To address this issue, this chapter is divided into three sections. The first will frame what is meant by International Energy Law, identifying not only its essential characteristics but also the main legal instruments and institutions that constitute this body of law. The second section explores the role of IEL in the context of the energy transition from both national and international perspectives. The third and final section delves into the potential of this subregime to address the particular challenges facing the energy transition in Latin America, in light of some intrinsic challenges of environmental, social, and economic governance in our region.

2 International Energy Law as an Emerging Discipline

2.1 *Early Understandings of Energy Law*

While the regulation of what we now identify as "energy resources" has been recognizable for centuries, specialized scholarship considers that the origins of Energy Law began with the regulation of working conditions for coal miners in the first half of the nineteenth century (Heffron & Talus, 2016). However, there is a relative consensus that the beginnings of Energy Law as a specific branch of the legal system are associated with the 1970s, in response to the so-called oil crisis and its main outcome: the wave of liberalization of energy companies, leading to the creation of a global energy market (Heffron et al., 2018).

Prior to that period, activities in the energy sector were typically managed exclusively by the State, where what is now recognized as Energy Law was indistinguishable from the special Administrative Law relevant to the management of public companies (Del Guayo Castiella, 2020). However, the oil crisis triggered a reconceptualization of energy from a national public utility to a matter of strategic and geopolitical significance.

Thus, the recognition of the finite nature of energy resources, along with the economic dependence they generated within States, consolidated the premise that the efficiency of the energy sector could be improved by introducing competition from private actors in energy generation (Heffron & Talus, 2016). Energy Law was therefore established from the outset as a framework of rules primarily aimed at creating and regulating energy markets, with a focus on national development.

It was not until the 1990s, during the consolidation of the *global energy market*, that Bradbrook (1996) coined one of the most frequently cited definitions of Energy Law—namely, "the allocation of rights and duties relating to the exploitation of all types of energy resources among individuals, between individuals and government, between governments, and between States."

Despite the delimitation offered by this concept, specialized scholarship was quick to criticize the "lack of purpose" in Energy Law as a regulatory regime (Heffron & Talus, 2016), which manifested in a lack of coherence in its philosophical and material development. While a full assessment of this shortcoming is beyond the scope of this work, it is known that Energy Law has been characterized by its fragmentation (Salimi, 2025) and its ambiguous impact on the deployment of energy infrastructure at the national and subnational levels.

Although classical doctrine still expresses some resistance to recognizing a "prescriptive" dimension of Energy Law, an increasing number of actors now understand this subset as the set of rules that we as a society have constructed, at multiple scales, to holistically govern the tensions arising in the generation, transport, and use of energy, given their significant environmental, social, and economic impacts—thus transcending the exclusive management of energy markets. This governance and its tensions are often expressed through the "energy trilemma" model (IEA, 2023) (Fig. 1).

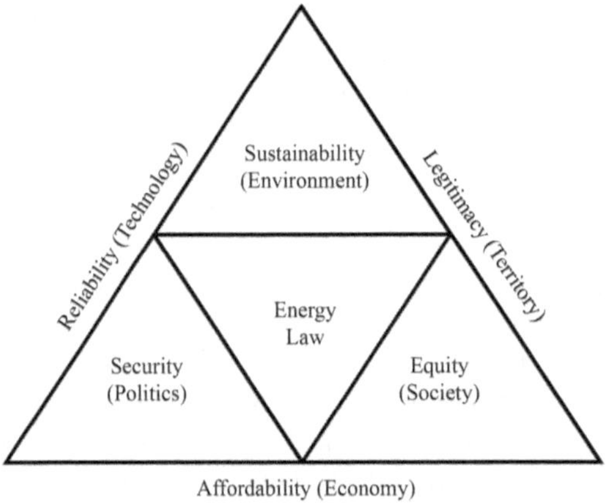

Fig. 1 Triangular framework for Energy Law in relation to the "energy trilemma", showing the intersection of Sustainability, Security, and Equity, framed by the external dimensions of Technology (Reliability), Territory (Legitimacy), and Economy (Affordability)

2.2 *The Internationalization of Energy Law*

Although the emergence of Energy Law is directly linked to the establishment of a global energy market, it is undeniable that this discipline originally laid its foundations as a regime of primarily national character. Not only was its implementation exclusively dependent on national laws and judicial decisions, but also for a long time it served national interests over global ones.

This framework is largely explained by a material variable: the energy sector, like other public utility systems, is particularly dependent on its infrastructure. In turn, infrastructure, as both the object and the condition of Energy Law, was designed and limited for many years in its capacity to meet national or subnational needs.

However, the proliferation of international agreements for the governance of energy-related matters—as well as environmental and economic issues with implications for the energy sector (such as the international governance of climate change)—has catalyzed a debate on the existence of IEL as a specific regime within International Law. Thus, rather than a phenomenon of "internationalization" of Energy Law, it is more accurate to speak of a process of *energization of International Law*.

Legal scholars who recognize IEL as a normative subset deserving special attention argue that its importance is not purely analytical; it also carries practical implications for the interpretation and application of both Energy Law and International Law, particularly regarding questions about the teleological dimension of the regime.

In this regard, Roeben and Mete (2020) identify two distinct types or levels of IEL. First, they recognize the existence of what could be translated as "International Law on Energy," which would include the set of international rules that belong to other sectors but that address fundamental aspects of energy systems' functioning. Additionally, the authors acknowledge the existence of "International Law of Energy" in the strict sense, which would consist of a series of international norms primarily dedicated to governing energy matters of an international nature.

The most relevant feature of the International Law of Energy—as a synonym of International Energy Law—is that it would be structured around a ***meta-norm*** mandating a "transition to a global low-carbon economy to provide access to sustainable and affordable energy for all" (Roeben & Mete, 2020). This meta-norm not only would determine the regulatory approach of IEL toward the global energy value chain but, more importantly, would also aim to prompt other regimes of International Law to ***realign themselves in accordance with this mandate.***

Because of the intrinsic characteristics of IEL, it would be virtually impossible to establish an exhaustive list of the normative instruments that comprise it. However, it is possible and useful to identify those instruments commonly included within this regime, in order to facilitate its understanding and framing.

To begin with, it is important to note that, unlike other branches of Public International Law such as Environmental Law or Maritime Law, IEL does not have a convention governing its fundamental aspects (Wawryk, 2014). Instead, legal scholars specializing in this discipline typically base their analysis on the list of sources of International Law established in Article 38 of the Statute of the International Court of Justice, which recognizes international conventions, international customs, and general principles of law (Nordtveit, 2020).

With regard to international conventions and treaties, there is a broad consensus that the multiple conventions regulating the development of nuclear energy are a source of IEL (Wawryk, 2014; Heffron, 2021), as are conventions dedicated to the recovery of the ozone layer and the mitigation of the negative impacts of climate change. More recently, some scholars have advocated—with relative success—for the inclusion of Human Rights conventions because of the implications and restrictions they impose on the development, transport, and use of energy resources (Hernández Palma et al., 2022).

Another important section of IEL is also embodied in bilateral treaties between States that regulate energy matters, with a notable prevalence of free trade agreements that directly or indirectly govern the trade of energy resources (Wawryk, 2014). Within this group, regional energy instruments developed by the European Union (EU) are often highlighted for their pioneering nature and their overall influence on the global energy market. On the other hand, in matters of customary law, the literature includes the so-called *Lex Petrolea*, which is applicable to the regulation of international trade and transport in the oil industry (Betin, 2022).

Regarding the principles of IEL, specialized scholarship has identified certain universal principles associated with Energy Law: the principle of security of supply, economic efficiency, and environmental sustainability, making clear reference to the energy trilemma. Additionally, there is recognition of a certain level of influence from principles of related branches. The principle of sovereignty over energy resources, adapted from the Rio Declaration on Environment and Development, is a paradigmatic example of this.

An instrument that combines or transcends these categories, and has therefore attracted special attention from specialized scholarship, is the Energy Charter Treaty (ECT). Thus, complemented by its respective "Protocol on Energy Efficiency and Related Environmental Aspects," the ECT constitutes one of the most important conventions for the consolidation of the global energy market.

Specifically, the ECT aims to "establish a legal framework to promote long-term cooperation in the energy sector ... [to] achieve an open and competitive market for energy materials and products." Thus, it contains the highest density of direct rules for international trade, transport, and investment in fossil fuels. Nevertheless, a significant portion of the literature does not recognize this instrument as the governing framework of IEL, emphasizing that its membership is not universal—comprising only 53 member States—and that it does not address all issues necessary for the international governance of energy, which extend far beyond mere trade (Roeben & Mete, 2020).

Beyond the debates regarding the content and characteristics of IEL, the analysis of this subfield has gained renewed prominence in recent years because of its expansion, driven by the catalyzation of international trade in "new" energy resources, including so-called critical minerals and low-carbon fuels. This new phase of IEL has sparked interest in the potential role that IEL may play in accelerating and steering this agenda toward specific outcomes and values.

3 The Role of International Energy Law in the Energy Transition

The marked expansion and diversification of IEL, in light of the now undeniable international nature of energy in a globalized world (Heffron, 2023), has not yet consolidated its importance as a special regime of International Law; rather, it has amplified the pre-existing levels of fragmentation and normative dispersion that have characterized this regime since its inception.

In this context, as global energy markets appear to be entering a second phase catalyzed by the globalization of the energy transition toward renewables, the exercise of characterizing IEL has transcended its theoretical–academic purpose and has become an endeavor with regulatory and public policy implications. Thus, it not only seeks to evaluate the coherence of regulations related to the energy transition but also becomes a potential tool to *steer this regulatory framework toward certain interests or outcomes* by strengthening specific values or principles.

In this regard, while it is reasonable to assume that all possible values shaping IEL lead toward more comprehensive and effective governance of the energy trilemma, the reality is that—just as with the well-known dimensions of sustainability (namely, environmental, social, and economic)—it would be naïve to think that the dimensions of the energy trilemma are in perfect harmony and balance in their weighting for energy transition decision-making. In fact, it is increasingly evident that there is a contest over the prioritization of the different dimensions of the trilemma in which energy security seems to have been prioritized in various legal systems (Fig. 2).

In this context, the analysis of the potential role of IEL in accelerating, guiding, and controlling the energy transition necessarily involves answering a prior question regarding the existence of an ethos that provides coherence to the development of this subfield. In other words, to identify the potential of IEL, it is necessary to ask whether the development of its

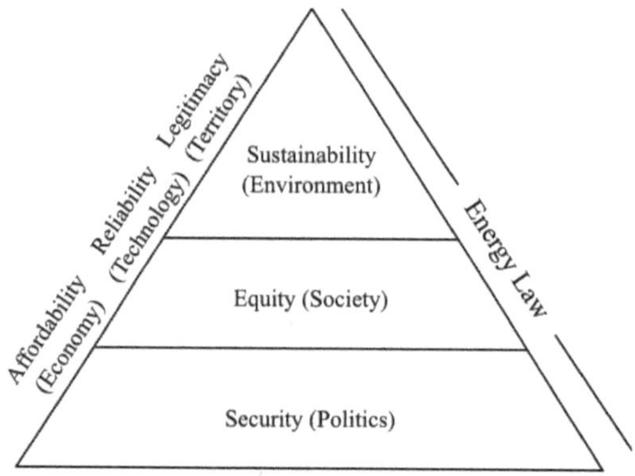

Fig. 2 Alternative Pyramid framework of Energy Law in relation to a reintepretation of the "energy trilemma". The *internal tiers* stack Sustainability (*top*), Equity (*middle*), and Security (*bottom*). The *left exterior edge* lists Affordability, Reliability, and Legitimacy, while the *right edge* shows Energy Law

instruments responds—and to what extent—to the existence of a general mandate that could be discerned from the main instruments comprising this special regime of International Law.

With this in mind, it is prudent to return to the theory proposed by Roeben and Mete (2020) on the acknowledgment of a meta-norm in IEL. According to the authors, there is agreement within the international community on the need and feasibility of modifying the energy matrix, at all scales, to provide access to sustainable and affordable energy for all. Considering this to be true, there is a ***meta-norm toward a sustainable energy transition***, as an imperative premise at the core of modern IEL, which would condition the creation of new instruments and organizations operating within the regime's frameworks. More importantly, this mandate would also drive the realignment of other International Law tools established prior to this consensus.

Although the literature acknowledges that this meta-norm is not ontological per se, but is identified through the "study of actors' behavior with respect to social rules in complex governance systems" (Roeben & Mete, 2020), it is common for such mandates to eventually be formalized through resolutions of the international community. Thus, some authors do recognize the adoption of Sustainable Development Goal 7 (SDG 7) by the United Nations General Assembly—on sustainable and affordable energy—as a milestone or foundational moment, subsequently reinforced by the adoption of the Paris Agreement (Dzebo et al., 2019).

Given its characteristics, the existence and effectiveness of a meta-norm is often demonstrated by its capacity to condition the creation of new instruments or to realign those already in force. In this regard, perhaps the most emblematic example, alongside the development of the European Green Deal and other EU energy strategies, is the case of the Energy Charter Treaty (ECT). This treaty underwent a formal "modernization" process in response to a request from certain States, multilateral organizations, and civil society actors for its realignment with the new commitments established by the Paris Agreement (Cocciolo & Reins, 2024). Specifically, it requested the modification of investor–state dispute settlement (ISDS) mechanisms, which have been widely used to oppose, through international arbitration, the implementation of decarbonization policies by member States (Crossley, 2019).

Despite the transformational nature of this event, it is still difficult to classify the ECT realignment process as successful. In 2022, after nearly a decade of negotiations, a preliminary agreement was reached that was deemed "unacceptable" by multiple stakeholders, as it protected existing investments in fossil fuels until 2033 and did not consider changes to the dispute settlement mechanisms. The outcome generated such rejection that from that year onward, a wave of departures by several emblematic ECT members was observed—including Spain, the Netherlands, France, and Germany—citing the inconsistency of the dispute settlement mechanism with the Paris Agreement and the European Green Deal. Two years later, both the EU and the United Kingdom announced their coordinated withdrawal from the ECT, which has been interpreted by many as the effective end of the treaty for all practical purposes.

Therefore, far from being a case of historic realignment, the ECT has become a parable of what happens to instruments of Public International Law that resist the advance of the meta-norm. A more recent example, whose development and impact remain to be seen, is the First International Conference for the Phasing Out of Fossil Fuels, to be led by Colombia and the Netherlands, and announced at the 30th Conference of the Parties to the United Nations Framework Convention on Climate Change (COP30) in Belém, as a response to the lack of clear references to fossil fuel phase-out in the decisions of the Conference of the Parties to the Paris Agreement.

Having thus established the potential of IEL to realign instruments of International Law—and, by extension, the existence of a meta-norm or general mandate guiding its development—to project the impact of this regime in the ongoing energy transition, we need to understand the roles that researchers have recognized for Energy Law in relation to the energy transition (Huhta, 2022), assessing its potential extrapolation to IEL.

It has been identified that Energy Law can play a role as (1) a *facilitator of the energy transition* through the recognition of realities—and fictions—that help mitigate the costs and risks of certain strategies. On the other hand, another potential role is recognized as (2) a *restrictor of the energy transition*, limiting or prohibiting certain technologies, as has occurred with nuclear energy. Finally, a complementary role is also included as (3) a *driver of the energy transition*, taking a more prominent role in defining goals and objectives, as well as in the mechanisms for their monitoring and evaluation.

In this regard, there is a relative consensus that, at least over the past decade, IEL has played a facilitating role, proposing mechanisms to promote the competitiveness of renewable energies over fossil fuels, and now incorporating measures to facilitate trade in critical minerals and low-carbon fuels. However, the reality is that prior to the consolidation of the meta-norm—marked by the establishment of the Paris Agreement and SDG 7—Energy Law was mainly used to shield and ensure the functioning of the fossil fuel regime, with one of the main examples being the ECT itself and its ISDS mechanism.

In this sense, a significant part of the existing skepticism regarding IEL as a coherent and effective regime relates to its inability to apply its restrictive potential to limit the expansion of fossil fuels. Thus, despite the consensus reached in the Paris Agreement and its deepening during the Global Stocktake process of the Agreement—where, for the first time, the need to "transition away from fossil fuels in energy systems, in a just, orderly and equitable manner" (UNFCCC, 2023) was affirmed—the processes of realigning regulations and policies outside the European context have been extremely limited.

Given this scenario of IEL impotence regarding the phase-out of fossil fuels, it would seem that the potential to strengthen the restrictive role of IEL is directly linked to the inclusion of Human Rights instruments as part of this subset. In fact, a framework that has stood out—for better or worse—in its implications for the global energy transition is the Human Rights and Business framework, where the EU remains the main reference, and which has led to the adoption of high-impact instruments, such as the

Corporate Sustainability Due Diligence Directive (CSDDD). However, precisely because of its transformative potential, which imposes requirements on companies, these have been countered early on through *legislative simplification packages*—"Omnibus" I and II—aimed at reducing regulatory burdens, improving sustainability standards, and unlocking investment.

It is likely that these fluctuations in the application of IEL, and the apparent resistance to the inclusion of Human Rights frameworks as relevant elements of this regime, are in fact an inevitable part of realignment processes. However, the reality is that the eventual failure of these processes—which we will only be able to identify as we approach the end of this critical decade—seriously jeopardizes the legitimacy of the instruments that consolidate the meta-norm and, with it, the existence of IEL and other International Law subregimes as coherent and effective tools. Nevertheless, it appears that at the regional level, IEL still expresses itself in its role as a driver, at least in the European context. For this reason, its effectiveness in other regions, such as Latin America, could be an important strategy to give this regulatory framework a second wind.

4 The Potential of International Energy Law for the Energy Transition in Latin America

As noted in the previous section, IEL has played various roles in the development of the global energy transition, primarily serving as a facilitator of energy resource trade, enabling strategies to reduce the costs of phasing out fossil fuels and phasing in renewable energies. It was also identified that while it is possible to recognize a meta-norm aimed at ensuring *access to sustainable and affordable energy for all*, the developments within this subsystem have not been sufficient to fully realize this objective. However, rather than a case of total ineffectiveness, it appears that IEL has been instrumentalized to prioritize and reinterpret certain dimensions of its mandate, transforming the mandate of accessibility into one of national security and economic sustainability of energy systems.

The situation in Latin America is no different. As a region highly dependent on the global energy market—holding around 15% of the world's oil and natural gas resources—most of the IEL applicable to Latin America concerns the trade of fossil fuels. Only very recently have some trade agreements with other regions been included to enable an energy transition, mainly expressed in agreements to exploit and export so-called critical minerals, as well as the commercialization of low-carbon fuels and industrial products.

On the other hand, the situation in Latin America is substantially different from what it was just a decade ago. The region's emergence as a leader in renewable energy development has not resulted in a rapid transformation of its energy matrix, but it has triggered a literal race between States to lead the supply of the global energy transition market. Thus, in addition to the development of critical minerals extraction projects and the establishment of international cooperation agreements for the generation of energy carriers such as green hydrogen (H2V) or green ammonia, countries like Chile, Brazil, and others are projecting the catalyzation of transatlantic trade in energy transition commodities on a scale that, until a few years ago, seemed technically and politically impossible.

In summary, a combination of factors such as the decreasing cost of renewable energy technologies (particularly wind and photovoltaic energy (Roser, 2020), the prioritization of an energy security approach in Europe as a result of the economic and energy impacts of the Russia–Ukraine war (Kuzemko et al., 2022), and the geopolitical pressures from China as the main producer of minerals and technologies for renewable energy generation have led to a series of regulatory innovations aimed at seizing a supposed window of opportunity for Latin American States. Thus, one of the main expressions of IEL in Latin America has been its role as a facilitator of political, technical, and economic variables for an "accelerated energy transition" (Newell et al., 2022).

However, the application of IEL to implement this political–economic agenda can easily be characterized as a distortion. This is because the agenda promoted by IEL in Latin America has limited implications for transforming the energy matrix toward clean and secure energies, dismissing the fulfillment of its meta-norm in this region, in contrast with the application of IEL in the European context. In fact, there is a—potentially intentional—confusion between energy transition strategies and "green industrialization" strategies in Latin America (Climate Tracker, 2024), which, while often related, can also involve significant contradictions when implemented simultaneously.

If we take the case of lithium, recently classified as a critical mineral because of its importance for electromobility and energy storage, it can be stated with certainty that the extraction and export of lithium does not directly contribute to the energy transition of the country extracting it, unless there are additional international cooperation agreements between the countries involved in the transaction. In this sense, a country could extract and sell all of its lithium reserves without this contributing at all to reducing its emissions or improving its levels of energy access.

On the contrary, the supplying country would likely experience higher emissions and a considerably reduced adaptive capacity in its salt flats and linked ecosystems—including, of course, the local communities that depend on them to survive and thrive. Despite this, it is common for countries that are part of the so-called lithium triangle to frame their lithium extraction agendas within their energy transition strategies.

A similar situation occurs with the projections for the green hydrogen industry, with some important differences. Latin American States seeking to position themselves as exporters of green hydrogen, as a potential vector for the trade and transport of renewable energies, often consider allocating a certain percentage of the hydrogen that is produced for use in decarbonizing non-electrifiable industries within their national territory. Thus, its development does present some potential to contribute to the energy transition and carbon neutrality goals of certain industrial sectors in the region. However, when this dynamic is analyzed at the subnational level, it is common to find proposals for "hydrogen valleys" where production is projected exclusively for export. In such cases, the local population does not perceive any change in the functioning of their energy systems.

In the best-case scenario, the effect of IEL in Latin America has been to catalyze the proliferation of legislation, regulation, and policy on energy transition. In only a few cases—such as Chile and Brazil—has this been accompanied by certain improvements in renewable energy capacity indices (World Economic Forum, 2024), although it remains difficult to establish a causal link between these processes. Indeed, many of the indicative policies on energy transition, rather than enabling new strategies or technologies, seem primarily intended to send a political message to other States or investors about the existence of a "favorable environment" for investment in these areas (Podesta et al., 2022).

Therefore, although there is a prevailing notion that economic growth and alliances forged through trade relations to enable green industrialization in the region will eventually facilitate the necessary trade flows for the region's energy transition, the lack of certainty regarding their impact on the energy matrix of Latin American countries calls into question even their classification as instruments of IEL, especially if we recognize that such instruments should be aligned with the meta-norm of a sustainable and affordable energy transition for all.

A prominent example of such instruments of questionable classification as IEL instruments is the *EU Global Gateway Investment Agenda*, which has been defined as a geopolitical instrument for sustainable, long-term infrastructure investment in EU partner countries, in line with the G7 countries' commitment to develop collaborations to achieve global infrastructure development goals consistent with the 2030 Agenda and the Paris Agreement. Specifically, the initiative involves the mobilization of €300 billion in investments for "sustainable projects," with a significant portion allocated to Latin America to explicitly "speeding up a fair green transition" (European Commission, 2025). In parallel, under similar contexts and rationales, new consensuses on climate action and energy transition have driven the review and modernization of certain Advanced Framework Agreements (AFAs) between the EU and some countries in the region.

Beyond the evident efforts of the EU to consolidate Latin America as an ally and supplier within the framework of its own energy transition processes—a dynamic that some scholars have described as an expression of "energy colonialism" (Sánchez et al., 2024)—there are certain endogenous initiatives in our region that can be considered realignment processes in response to developments in IEL. An example of this is the regional "Renewables in Latin America and the Caribbean" initiative coordinated by the Inter-American Development Bank, which seeks, through voluntary agreements, to achieve a 70% share of renewables in the region's energy matrix by 2030 and already has the endorsement of 15 countries.

Similarly, the Latin American Energy Organization (OLADE), a regional multilateral body with 27 member countries, has focused on signing cooperation agreements with other regional and international actors to facilitate spaces for energy transition, despite facing strong criticism for considering natural gas as a "transition fuel."

Finally—and closely related to the point raised at the end of the previous section—perhaps the most significant potential for the region, in light of the socioenvironmental impacts projected by energy transition and green industrialization agendas, lies in the development of international Human Rights instruments and their potential application as constraints or drivers of certain energy transition processes. For example, the Regional Agreement on Access to Information, Public Participation, and Access to Justice in Environmental Matters—better known as the Escazú Agreement—has emerged as a tool with the potential to establish safeguards in phase-out processes for fossil fuels and phase-in processes for renewables, in line with the meta-norm or general mandate underlying IEL.

5 Conclusions

Despite the potential of IEL to catalyze the realignment of energy regulation and policy, and the opportunity offered by the explicit integration of international Human Rights instruments as drivers in energy transition processes, the reality is that the diagnosis regarding the limitations of IEL in its restrictive dimension—curbing the expansion of fossil fuel extraction and combustion—and in steering truly sustainable and affordable energy transition processes has not just been repeated in Latin America but also has been particularly pronounced.

Echoing the global trend, the vast majority of international hard and soft law instruments impacting the development of the Latin American energy transition do so from an energy security perspective, neglecting the affordability and sustainability factors required for a transformation of our energy matrix.

At a time of restructuring in the global energy market, where various economic powers are at the peak of their economic and political deployment to secure their own energy security, Latin American States have been tempted to embark on a "race to the bottom" in regulatory terms (Crossley, 2019), aiming to establish themselves as strategic suppliers in a new renewable energy market.

It is precisely in this context, where the energy transition appears to be shifting from internationalism to nationalism (Talus, 2024), that IEL can play a fundamental role. Thus, strengthening of IEL, guided by a meta-norm towards climate justice and a sustainable energy transition, has become an essential strategy to consolidate the progress achieved to date and to aim for a "race to the top" in the energy legal systems of our region —a race that may have many routes, but where only one will lead us with certainty to the goal of a prosperous future.

References

Betin, B. (2022). Lex petrolea in international arbitration - a very english dissent. *Journal of World Energy Law and Business*, *15*, 4. https://doi.org/10.1093/jwelb/jwac016

Bradbrook, A. J. (1996). Energy law as an academic discipline. *Journal of Energy & Natural Resources Law*, *14*(2). https://doi.org/10.1080/02646811.1996.11433062

Climate Tracker. (2024). *Green industrialization and new development models in Brazil, Colombia, and Chile: An introduction for journalists.* https://climatetrackerlatam.org/herramientas/guia-procesos-de-industrializacion-verde/

Cocciolo, E., & Reins, L. (2024). A critical review of the energy charter treaty from an earth system law perspective. *Transnational Environmental Law*, 1–27. https://doi.org/10.1017/S2047102524000244

Crossley, P. (2019). *Renewable energy law: An international assessment.* Cambridge University Press.

Del Guayo Castiella, Í. (2020). Concept, contents, and principles of energy law. *Revista de Administración Pública*, *212*, 309–345. https://doi.org/10.18042/cepc/rap.212.12

European Commission (2025). *EU-LAC Global Gateway investment agenda.* International Parnerships. https://international-partnerships.ec.europa.eu/policies/global-gateway/eu-lac-global-gateway-investment-agenda_en

Kuzemko, C., Blondeel, M., Dupont, C., & Brisbois, M. C. (2022). Russia's war on Ukraine, European energy policy responses & implications for sustainable transformations. *Energy Research and Social Science*, *93*, https://doi.org/10.1016/j.erss.2022.102842

Dzebo, A., Janetschek, H., Brandi, C., & Iacobuta, G. (2019). *Connections between the Paris agreement and the 2030 Agenda: The case for policy coherence.* Working Paper, Stockholm Environment Institute. https://www.sei.org/wp-content/uploads/2019/08/connections-between-the-paris-agreement-and-the-2030-agenda.pdf

Heffron, R. J., & Talus, K. (2016). The evolution of energy law and energy jurisprudence: Insights for energy analysts and researchers. *Energy Research and Social Science*, *19*, https://doi.org/10.1016/j.erss.2016.05.004

Heffron, R. J., Rønne, A., Tomain, J. P., Bradbrook, A., & Talus, K. (2018). A treatise for energy law. *Journal of World Energy Law and Business*, 11. https://doi.org/10.1093/jwelb/jwx039

Heffron, R. J. (2021). *Energy law: An introduction* (2 ed.). Springer Briefs in Law.

Heffron, R. J. (2023). Energy law for the next generation, towards 2030 to 2050. *Journal of Energy & Natural Resources Law*, *41*(2). https://doi.org/10.1080/02646811.2023.2190688

Hernández Palma, H. G., Jiménez Coronado, A., & Mendoza Casseres, D. (2022). Renewable energy and international energy law. *Justicia, 27*(41). https://doi.org/10.17081/just.27.41.5997

Huhta, K. (2022). The contribution of energy law to the energy transition and energy research. *Global Environmental Change, 73*, https://doi.org/10.1016/j.gloenvcha.2021.102454

International Energy Agency. (2023). *Overcoming the energy trilemma: Secure and inclusive transitions. IEA report to G7 Leaders.* https://www.iea.org/reports/overcoming-the-energy-trilemma-secure-and-inclusive-transitions

Newell, P. J., Geels, F. W., & Sovacool, B. K. (2022). Navigating tensions between rapid and just low-carbon transitions. *Environmental Research Letters, 17*(4), 041006. https://doi.org/10.1088/1748-9326/ac622a

Nordtveit, E. (2020). International energy law in perspective: The relationship between national and international energy law. *Routledge Handbook of Energy Law.*

Podesta, A., Eirin, M. S., Contreras, R., & Salgado, R. (2022). Investment attraction policies for financing clean energy in Latin America. In *Project documents, economic commission for Latin America and the caribbean (ECLAC).*

Roser, M. (2020). *Why did renewables become so cheap so fast Our World in Data.* https://ourworldindata.org/cheaprenewables-growth

Roeben, V., & Mete, G. (2020). What do we mean when we talk about international energy law? *The global energy transition: law, policy and economics for energy in the twenty-first century.* Hart Publishing. https://doi.org/10.5040/9781509932511.ch-004

Salimi, H. (2025). Fragmentation in international energy law: Managing an inevitable phenomenon. *International Journal of Public Law and Policy, 1*(1). https://doi.org/10.1504/IJPLAP.2023.10058002

Sánchez, J., Matarán, A., Campos-Celador, A., & Fjellheim, E. M. (2024). Energy colonialism: A category to analyse the corporate energy transition in the global South and North. *Land, 12*, 1241. https://doi.org/10.3390/land12061241

Talus, K. (2024). *The disciplinary foundations of energy law: In a research agenda for international energy law.* Edward Elgar Publishing. https://doi.org/10.4337/9781803924960

UNFCCC (2023). Report of the Conference of the Parties serving as the meeting of the Parties to the Paris Agreement on its fifth session, held in the United Arab Emirates from 30 November to 13 December 2023 (Decision 1/CMA.5). https://unfccc.int/topics/global-stocktake/about-the-global-stocktake/outcome-of-the-first-global-stocktake

Wawryk, A. S. (2014). *International energy law: An emerging academic discipline.* Adelaide Law School Research Paper No. 2014–16.

World Economic Forum. (2024). *Fostering effective transition.* Insight report. https://www3.weforum.org/docs/WEF_Fostering_Effective_Energy_Transition_2024.pdf

FSC
www.fsc.org
MIX
Papier | Fördert gute Waldnutzung
FSC® C083411

Kolibri 360 GmbH
Ferdinand-Jühlke-Straße 7
99095 Erfurt, Deutschland
produktsicherheit@kolibri360.de